Your Role in the Green Environment

Trainee Guide
Third Edition, LEED v4

PEARSON

Boston   Columbus   Indianapolis   New York   San Francisco   Amsterdam
Cape Town Dubai London Madrid Milan Munich Paris Montreal Toronto Delhi
Mexico City   Sao Paulo   Sydney   Hong Kong   Seoul   Singapore   Taipei   Tokyo

NCCER

President: Don Whyte
Director of Product Development: Daniele Dixon
Your Role in the Green Environment Project Manager:
 Jamie Carroll
Senior Manager of Production: Tim Davis

Quality Assurance Coordinator: Debie Hicks
Desktop Publishing Coordinator: James McKay
Permissions Specialist: Adrienne Payne
Production Specialist: Adrienne Payne
Editor: Tanner Yea

Author:

Dr. Annie Pearce

Development services provided by Topaz Publications, Liverpool, NY

Project Manager: Veronica Westfall
Desktop Publisher: Joanne Hart

Art Director: Alison Richmond
Permissions Editor: Tonia Burke

Pearson Education, Inc.

Director, Global Employability Solutions: Jonell Sanchez
Head of Associations: Andrew Taylor
Editorial Assistant: Kelsey Kissner
Program Manager: Alexandrina B. Wolf
Project Manager: Janet Portisch
Operations Supervisor: Deidra M. Skahill
Art Director: Diane Ernsberger
Digital Product Strategy Manager: Maria Anaya
Digital Studio Project Managers: Heather Darby,
 Tanika Henderson
Directors of Marketing: David Gesell, Margaret Waples
Field Marketer: Brian Hoehl

Composition: NCCER
Printer/Binder: Courier/Kendallville
Cover Printer: Courier/Kendallville
Text Fonts: Palatino and Univers

Credits and acknowledgments for content borrowed from other sources and reproduced, with permission, appear at the end of this textbook.

10 9 8 7 6 5 4 3 2 1

PEARSON

Paperback: ISBN 13: 978-0-13-294863-0
ISBN 10: 0-13-294863-X

Preface

To the Trainee

If you're just getting started on the path towards a career in construction, you'll soon learn about the tools you'll need to carry you forward—from safety awareness, to construction math, to working with hand and power tools. You'll also discover how much these skills can help you on the job, and how well they can transfer into your everyday life. While tackling that next home project, you'll find your new skills will come in handy—enabling you to manage the project more efficiently and safely than you would have before.

You'll not only learn how much of an impact construction has on the environment and ways in which you can reduce the impact, but you'll also learn about the enormous influence you can have on the environment in your day-to-day life. While these skills may not directly affect how well you accomplish that next home project, you'll likely approach it with a deeper understanding than you had before.

New with this Revision of *Your Role in the Green Environment*

This revision to *Your Role in the Green Environment* contains the following enhancements:

- All LEED content has been updated to reflect LEED Version 4.0, with a focus on standards for Building Design + Construction. For more information on the latest version of LEED, visit the US Green Building Council at **www.usgbc.org**.
- Discussion of safety issues and challenges for green buildings are placed throughout the text, including new safety warnings.
- There are new discussions/feature boxes on tradeoffs in green buildings, including potential tradeoffs between environmental benefits and human health and safety, green products and product durability or availability, and others.
- New features on contemporary issues including net zero buildings and building automation have been added.
- All worksheets and data are provided in both US and metric units.
- An expanded focus now includes issues relevant for international construction.
- Statistics have been updated throughout.

Successful completion of this course could provide you with up to 15 continuing education hours towards the maintenance of your green building credential (i.e. LEED AP and/or LEED Green Associate). For details, visit **http://www.gbci.org/CMP/about-cmp.aspx**.

We invite you to visit the NCCER website at **www.nccer.org** for information on the latest product releases and training, as well as online versions of the *Cornerstone* magazine and Pearson's NCCER product catalog.

Your feedback is welcome. You may email your comments to **curriculum@nccer.org** or send general comments and inquiries to **info@nccer.org**.

NCCER Standardized Curricula

NCCER is a not-for-profit 501(c)(3) education foundation established in 1995 by the world's largest and most progressive construction companies and national construction associations. It was founded to address the severe workforce shortage facing the industry and to develop a standardized training process and curricula. Today, NCCER is supported by hundreds of leading construction and maintenance companies, manufacturers, and national associations. The NCCER Standardized Curricula was developed by NCCER in partnership with Pearson Education, Inc., the world's largest educational publisher.

Some features of the NCCER Standardized Curricula are as follows:

- An industry-proven record of success
- Curricula developed by the industry for the industry
- National standardization providing portability of learned job skills and educational credits
- Compliance with Office of Apprenticeship requirements for related classroom training (*CFR 29:29*)
- Well-illustrated, up-to-date, and practical information

NCCER also maintains a Registry that provides transcripts, certificates, and wallet cards to individuals who have successfully completed a level of training within a craft in NCCER's Standardized Curricula. *Training programs must be delivered by an NCCER Accredited Training Sponsor in order to receive these credentials.*

Special Features

In an effort to provide a comprehensive user-friendly training resource, we have incorporated many different features for your use. Whether you are a visual or hands-on learner, this book will provide you with the proper tools to get started in green construction.

Introduction Page

This page lists the Objectives, Performance Tasks, and Trade Terms for this textbook. The Objectives list the skills and knowledge you will need in order to complete this training successfully. The Performance Tasks give you an opportunity to apply your knowledge to real-world tasks. The list of Trade Terms identifies important terms you will need to know.

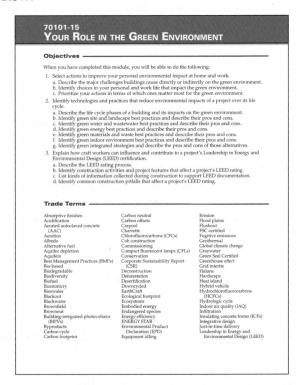

Special Features

Features present technical tips and professional practices. These features presentreal-life scenarios similar to those you might encounter on the job site.

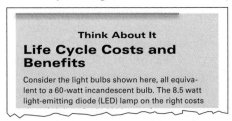

Notes, Cautions, and Warnings

Safety features are set off from the main text in highlighted boxes and organized into three categories based on the potential danger of the issue being addressed. Notes simply provide additional information on the topic area. Cautions alert you of a danger that does not present potential injury but may cause damage to equipment. Warnings stress a potentially dangerous situation that may cause injury to you or a co-worker.

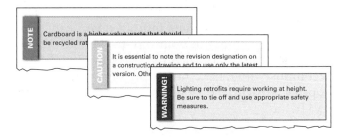

Color Illustrations and Photographs

Full-color illustrations and photographs are used throughout each module to provide vivid detail. These figures highlight important concepts from the text and provide clarity for complex instructions. Each figure is denoted in the text in *italic type* for easy reference.

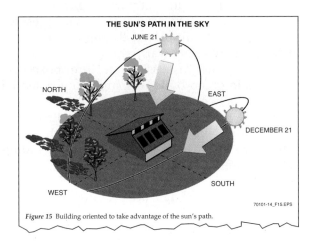

Figure 15 Building oriented to take advantage of the sun's path.

Going Green

Going Green looks at ways to preserve the environment, save energy, and make good choices regarding the health of the planet. Through the introduction of new construction practices and products, you will see how the "greening of the world" has already taken root.

Footprint Calculators

To calculate your ecological footprint, go to **www.myfootprint.org**. An online carbon footprint calculator is available at **www.carbonfootprint.com**. You can calculate your water footprint at **www.waterfootprint.org**. To see how much your own footprint equals in terms of impacts, go to **www.**

Did You Know?

The Did You Know? features introduce historical tidbits or modern information about the construction industry. Interesting and sometimes surprising facts about green construction are also presented.

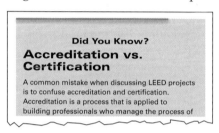

Did You Know?

Accreditation vs. Certification

A common mistake when discussing LEED projects is to confuse accreditation and certification. Accreditation is a process that is applied to building professionals who manage the process of

What's Wrong with this Picture?

What's Wrong with This Picture? features include photos of actual code violations for identification and encourage you to approach each installation with a critical eye.

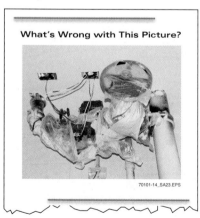

What's Wrong with This Picture?

70101-14_SA23.EPS

Review Questions

Review Questions are provided to reinforce the knowledge you have gained. This makes them a useful tool for measuring what you have learned.

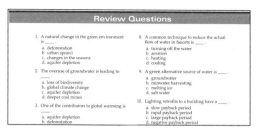

Review Questions

1. A natural change in the green environment is ____.
   a. deforestation
   b. urban sprawl
   c. changes in the seasons
   d. aquifer depletion

2. The overuse of groundwater is leading to ____.
   a. loss of biodiversity
   b. global climate change
   c. aquifer depletion
   d. deeper coal mines

3. One of the contributors to global warming is ____.
   a. aquifer depletion
   b. deforestation

8. A common technique to reduce the actual flow of water in faucets is ____.
   a. turning off the water
   b. aeration
   c. heating
   d. cooling

9. A green alternative source of water is ____.
   a. groundwater
   b. rainwater harvesting
   c. melting ice
   d. salt water

10. Lighting retrofits to a building have a ____.
   a. slow payback period
   b. rapid payback period
   c. large payback period
   d. negative payback period

Trade Terms

Trade Terms are denoted in the text with **bold blue type** upon their first occurrence. To make searches for key information easier, a comprehensive Glossary of Trade Terms is found at the back of this book.

The materials and processes used to construct a facility have the potential to create problems. Many modern finishes such as paints, sealants, carpets, and composites contain solvents and adhesives that release VOCs as they age. The off-gassing of VOCs is an example of fugitive emissions. These emissions come from the material

Sustainable Construction Supervisor

For those interested in pursuing advanced study, NCCER's *Sustainable Construction Supervisor*, endorsed by the Green Building Certification Institute (GBCI), will equip you to both recognize and implement green building practices.

CONTREN® LEARNING SERIES

Sustainable Construction Supervisor

GREEN BUILDING CERTIFICATION INSTITUTE

EDUCATION PROVIDER

TRAINEE GUIDE

NATIONAL CENTER FOR CONSTRUCTION EDUCATION AND RESEARCH

NCCERconnect

One Industry. One Training Program. One Online Solution.

NCCERconnect: An Interactive Online Course

Ideal for blended or distance education, NCCERconnect is a unique web-based supplement in the form of an electronic book that provides a range of visual, auditory, and interactive elements to enhance your training. It can be used in a variety of settings such as self-study, blended/distance education, or in the traditional classroom environment! It's the perfect way to review content from a class you may have missed or to practice at your own pace.

Features:

- **Online Lectures** – Each ebook module features a written summary of key content accompanied by an optional Audio Summary so if you need a refresher, this tool is always available.
- **Video Presentations** – Throughout, you'll find dynamic video presentations that demonstrate difficult skills and concepts! Of special note are the safe/unsafe scenarios shot on a live construction site testing your knowledge of the four 'high hazards' presented in the Basic Safety module.

- **Personalization Tools** – With the "highlighter" and "notes" options you can easily personalize your own NCCERconnect ebook to keep track of important information or create your own study guide.
- **Review Quizzes** – Short multiple-choice concept check quizzes at the end of each module section act as the ideal study tool and provide immediate feedback. Additionally, you'll find fill-in-the-blank trade terms quizzes, applied math questions, and comprehension questions at the end of each module.
- **Active Figures** – Interactive exercises bring key concepts to life, including animation in the Introduction to Construction Drawings module that will help you make the mental transition from a flat, 2-dimensional plan to a 3-dimensional finished structure.

Visit **www.nccerconnect.com** to view a demo. **NCCERconnect is available with:**

- Core Curriculum
- Carpentry Levels 1-4
- Construction Technology
- Electrical Levels 1-4
- Electronic Systems Technician Levels 1-4
- HVAC Levels 1-4
- Plumbing Levels 1-4
- Your Role in the Green Environment, LEED v3

Ask your Pearson Customer Service Representative for upcoming availability of *Your Role in the Green Environment*, LEED v4.

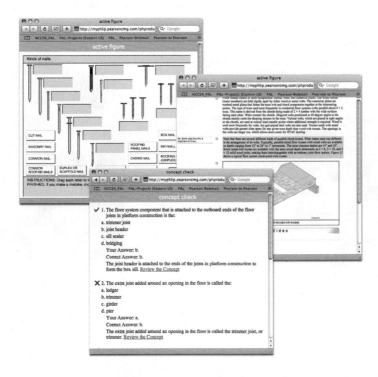

Foreword

The rapid growth of green building is important to everyone, and especially the construction industry workforce that is responsible for ensuring that buildings and infrastructure are being constructed according to a new set of requirements that includes consideration of both humans and nature. While some of the requirements of green building are very technical and require advanced knowledge of science and engineering, most of them are quite simple and obvious to most of us. We all know that we should not waste; we heard this from the time we were very young, and we continue to instruct our children about this concept. This way of thinking applies not only to our homes, but to everywhere and to everyone. Eliminating waste is one of the core principles of green building. Anyone who has worked in construction knows that there are typically enormous quantities of construction and demolition waste. Prior to the advent of green building this was thought of as being normal, but waste is nothing more than a sign of inefficiency. It makes little sense that products and materials made in factories from materials extracted from nature should end up being disposed of in engineered landfills that consume precious land and can threaten the quality of life for those who live in their proximity. Green building seeks to eradicate waste, keep resources in productive use, and close materials loops. Similarly, the effort of green construction addresses a number of other important, interrelated issues such as energy and water consumption, the waste of precious land resources, and the destruction of ecosystems. None of these critical resources should be wasted, as we owe it to our children and future generations to use them wisely. As a result, including green concepts in construction training is needed to give workers the information they need to change the process to one that is responsible and sustainable.

Training the construction workforce in the concepts of green building is already having positive benefits. There are currently examples of projects where over 95 percent of the waste that would have normally been generated was avoided through waste reduction efforts and diversion into continued productive uses. Similarly, the proper installation of systems, materials, and finishes, from glues and paint to ceiling tile, flooring, and air handlers is being accomplished in a manner that also enhances the health of building occupants, improving their productivity and making the United States a more competitive society.

This revised volume represents the continuing efforts of NCCER to support green building training through the NCCER training program sponsor network. It represents the further development of advanced training for construction workers, because it means that a potentially enormous audience is being informed about high-performance green building and its demanding requirements. As importantly, it is helping embed green building practices in the global construction community, thus addressing the issues of climate change and high energy costs.

Dr. Annie Pearce at Virginia Tech brought together a wide and varied array of information into this new volume and her efforts deserve acknowledgement and recognition. This training program continues the process of developing more responsible construction practices that are sorely needed to meet the rapidly growing demand for green building, as well as a parallel demand for a well-trained workforce that can deliver it.

NCCER's foresight in supporting the development of this revised volume should also be gratefully acknowledged by the larger United States green building movement because of the enormous positive impact it will have in educating contractors, subcontractors, supervisors, and craftspeople alike, in the building of environmentally responsible facilities and homes.

Dr. Charles J. Kibert, P.E.
Holland Professor
Powell Center for Construction & Environment
University of Florida
Gainesville, Florida

NCCER Standardized Curricula

NCCER's training programs comprise more than 80 construction, maintenance, pipeline, and utility areas and include skills assessments, safety training, and management education.

Boilermaking
Cabinetmaking
Carpentry
Concrete Finishing
Construction Craft Laborer
Construction Technology
Core Curriculum: Introductory Craft Skills
Drywall
Electrical
Electronic Systems Technician
Heating, Ventilating, and Air Conditioning
Heavy Equipment Operations
Highway/Heavy Construction
Hydroblasting
Industrial Coating and Lining Application Specialist
Industrial Maintenance Electrical and Instrumentation Technician
Industrial Maintenance Mechanic
Instrumentation
Insulating
Ironworking
Masonry
Millwright
Mobile Crane Operations
Painting
Painting, Industrial
Pipefitting
Pipelayer
Plumbing
Reinforcing Ironwork
Rigging
Scaffolding
Sheet Metal
Signal Person
Site Layout
Sprinkler Fitting
Tower Crane Operator
Welding

Maritime

Maritime Industry Fundamentals
Maritime Pipefitting
Maritime Structural Fitter

Green/Sustainable Construction

Building Auditor
Fundamentals of Weatherization
Introduction to Weatherization
Sustainable Construction Supervisor
Weatherization Crew Chief
Weatherization Technician
Your Role in the Green Environment

Energy

Alternative Energy
Introduction to the Power Industry
Introduction to Solar Photovoltaics
Introduction to Wind Energy
Power Industry Fundamentals
Power Generation Maintenance Electrician
Power Generation I&C Maintenance Technician
Power Generation Maintenance Mechanic
Power Line Worker
Power Line Worker: Distribution
Power Line Worker: Substation
Power Line Worker: Transmission
Solar Photovoltaic Systems Installer
Wind Turbine Maintenance Technician

Pipeline

Control Center Operations, Liquid
Corrosion Control
Electrical and Instrumentation
Field Operations, Liquid
Field Operations, Gas
Maintenance
Mechanical

Safety

Field Safety
Safety Orientation
Safety Technology

Supplemental Titles

Applied Construction Math
Tools for Success

Management

Fundamentals of Crew Leadership
Project Management
Project Supervision

Spanish Titles

Acabado de concreto: nivel uno
Aislamiento: nivel uno
Albañilería: nivel uno
Andamios
Carpintería: Formas para carpintería, nivel tres
Currículo básico: habilidades introductorias del oficio
Electricidad: nivel uno
Herrería: nivel uno
Herrería de refuerzo: nivel uno
Instalación de rociadores: nivel uno
Instalación de tuberías: nivel uno
Instrumentación: nivel uno, nivel dos, nivel tres, nivel cuatro
Orientación de seguridad
Paneles de yeso: nivel uno
Seguridad de campo

Acknowledgments

This curriculum was revised as a result of the farsightedness and leadership of the following sponsors:

ABC Green Committee
ABC Northern California Chapter
Breaking Ground Contracting
Brazosport College

Flintco, LLC
Flint Energy Services, Inc.
Rinker School of Building Construction at the
    University of Florida

This curriculum would not exist were it not for the dedication and unselfish energy of those volunteers who served on the Authoring Team. A sincere thanks is extended to the following:

Cassandra Dillon
Terrell Hoagland

Roy Horton
Charles Kibert
Tom Magness

Mary Tappouni
Ravi Srinivasan

Sustainable Facilities and Infrastructure Laboratory

This module was developed by members of the Sustainable Facilities & Infrastructure Laboratory of the Myers-Lawson School of Construction at Virginia Tech. Contributors included Dr. Annie Pearce, Dr. Christine Fiori, Mr. Sushil Shenoy, Dr. Yong Han Ahn, and Mr. Pratik Doshi. To ask questions or learn more about the ideas presented in this module, contact **sustainablefacilities@vt.edu**.

US Green Building Council

Your Role in the Green Environment, Version 4 has been approved for 15 general continuing education hours under GBCI's Credential Maintenance Program. For more information, visit **www.usgbc.org**.

NCCER Partners

American Fire Sprinkler Association
Associated Builders and Contractors, Inc.
Associated General Contractors of America
Association for Career and Technical Education
Association for Skilled and Technical Sciences
Construction Industry Institute
Construction Users Roundtable
Construction Workforce Development Center
Design Build Institute of America
GSSC – Gulf States Shipbuilders Consortium
ISN
Manufacturing Institute
Mason Contractors Association of America
Merit Contractors Association of Canada
NACE International
National Association of Minority Contractors
National Association of Women in Construction
National Insulation Association
National Technical Honor Society
National Utility Contractors Association
NAWIC Education Foundation
North American Crane Bureau
North American Technician Excellence
Pearson

Pearson Qualifications International
Prov
SkillsUSA®
Steel Erectors Association of America
U.S. Army Corps of Engineers
University of Florida, M. E. Rinker School of
    Building Construction
Women Construction Owners & Executives,
    USA

Contents

Your Role in the Green Environment

Geared to entry-level craftworkers, *Your Role in the Green Environment* provides pertinent information concerning the green environment, construction practices, and building rating systems. This edition has been updated to reflect LEED v4 with emphasis on standards for building design and construction. The updated content features contemporary issues such as net zero buildings and an expanded focus on issues relevant to international construction.

Your Role in the Green Environment, LEED v4, Third Edition has been approved by GBCI for 15 hours of general continuing education to support LEED professionals (Module ID 70101-15; 12.5 Hours)

Green Building Certification Institute
Credential Maintenance Program

Your Role in the Green Environment has been approved by the US Green Building Council (USGBC) for 15 general Continuing Education (CE) Hours for LEED Professionals.

USGBC requires that LEED Green Associates earn 15 continuing education hours biennially (3 LEED-specific). LEED Accredited Professionals must earn 30 continuing education hours biennially (6 LEED-specific).

To report your CE hours with USGBC, go to USGBC's web site at **www.usgbc.org**. and select "Credentials" from the drop down list at the top right. The Credential Maintenance guide can be accessed here: **www.usgbc.org/resources/cmp-guide**. Within the credentials section of your usgbc.org account, click "Report CE hours" to self-report your educational activity. Select "education" at the top of the page and enter the following details on the course: *Your Role in the Green Environment*, GBCI course ID 0920002816, NCCER (**www.nccer.org**), 15 general continuing education hours (LEED AP BD+C).

Should you have any questions regarding reporting CE hours or the USGBC Credentialing Maintenance Program, please contact USGBC at 1-800-795-1747 or at **www.usgbc.org/contactus**.

70101-15

Your Role in the Green Environment

OVERVIEW

The construction industry is always changing. In this new era, the green environment is an important consideration. As a construction craft worker, you should know how daily activities at work and at home affect the green environment. With this knowledge, you can make smart choices to reduce your impact. This module explains how your daily choices make a difference. You will learn to measure your carbon footprint and reduce it. You will also learn how buildings affect the green environment and how green building rating systems work.

Trainee Guide

Trainees with successful module completions may be eligible for credentialing through the NCCER Registry. To learn more, go to **www.nccer.org** or contact us at **1.888.622.3720**. Our website has information on the latest product releases and training, as well as online versions of our *Cornerstone* magazine and Pearson's product catalog.

Your feedback is welcome. You may email your comments to **curriculum@nccer.org**, send general comments and inquiries to **info@nccer.org**, or fill in the User Update form at the back of this module.

This information is general in nature and intended for training purposes only. Actual performance of activities described in this manual requires compliance with all applicable operating, service, maintenance, and safety procedures under the direction of qualified personnel. References in this manual to patented or proprietary devices do not constitute a recommendation of their use.

Objectives

When you have completed this module, you will be able to do the following:

1. Select actions to improve your personal environmental impact at home and work.
   a. Describe the major challenges buildings cause directly or indirectly on the green environment.
   b. Identify choices in your personal and work life that impact the green environment.
   c. Prioritize your actions in terms of which ones matter most for the green environment.
2. Identify technologies and practices that reduce environmental impacts of a project over its life cycle.
   a. Describe the life cycle phases of a building and its impacts on the green environment.
   b. Identify green site and landscape best practices and describe their pros and cons.
   c. Identify green water and wastewater best practices and describe their pros and cons.
   d. Identify green energy best practices and describe their pros and cons.
   e. Identify green materials and waste best practices and describe their pros and cons.
   f. Identify green indoor environment best practices and describe their pros and cons.
   g. Identify green integrated strategies and describe the pros and cons of those alternatives.
3. Explain how craft workers can influence and contribute to a project's Leadership in Energy and Environmental Design (LEED) certification.
   a. Describe the LEED rating process.
   b. Identify construction activities and project features that affect a project's LEED rating.
   c. List kinds of information collected during construction to support LEED documentation.
   d. Identify common construction pitfalls that affect a project's LEED rating.

Performance Tasks

This is a knowledge-based module; there are no performance tasks.

Trade Terms

Absorptive finishes
Acidification
Aerated autoclaved concrete (AAC)
Aeration
Albedo
Alternative fuel
Aquifer depletion
Aquifers
Best Management Practices (BMPs)
Bio-based
Biodegradable
Biodiversity
Biofuel
Biomimicry
Bioswales

Blackout
Blackwater
Brownfield
Brownout
Building-integrated photovoltaics (BIPVs)
Byproducts
Carbon cycle
Carbon footprint
Carbon neutral
Carbon offsets
Carpool
Charrette
Chlorofluorocarbons (CFCs)
Cob construction
Commissioning

Compact fluorescent lamps (CFLs)
Conservation
Corporate Sustainability Report (CSR)
Deconstruction
Deforestation
Desertification
Downcycled
EarthCraft
Ecological footprint
Ecosystems
Embodied energy
Endangered species
Energy efficiency
ENERGY STAR

(continued)

Trade Terms (continued)

Environmental Product Declaration (EPD)
Equipment idling
Erosion
Flood plains
Flushout
FSC certified
Fugitive emissions
Geothermal
Global climate change
Graywater
Green Seal Certified
Greenhouse effect
Grid intertie
Halons
Hardscape
Heat island
Hybrid vehicle
Hydrochlorofluorocarbons (HCFCs)
Hydrologic cycle
Indoor air quality (IAQ)
Infiltration
Insulating concrete forms (ICFs)
Integrative design
Just-in-time delivery
Leadership in Energy and Environmental Design (LEED)
Life cycle
Life cycle assessment
Life cycle cost
Light-emitting diode (LED) lamp
Lithium ion (Li-ion)
Local materials
Location valuation
Low-emission vehicle
Maintainability

Massing
Minimum Efficiency Reporting Value (MERV)
Monofill
Multi-function materials
Nano-materials
Nickel cadmium (NiCad)
Nonrenewable
Nontoxic
Offgassing
Ozone depletion
Ozone hole
Papercrete
Passive solar design
Passive survivability
Pathogens
Payback period
Peak shaving
Persistent, bioaccumulative toxin (PBT)
Pervious concrete
Phantom loads
Phase change materials (PCMs)
Photosensor
Photovoltaics
Pollution prevention
Post-consumer
Post-industrial/pre-consumer
Preservation
Radio-frequency identification (RFID)
Rainwater harvesting
Rammed earth
Rapidly renewable
Raw materials
Recyclable
Recycled content

Recycled plastic lumber (RPL)
Recycling
Renewable
Reusable
Salvaged
Sedimentation
Sick Building Syndrome
Smart materials
Softscape
Solar
Solid waste
Solvent-based
Spoil pile
Sprawl
Stormwater runoff
Strawbale construction
Structural insulated panels (SIPs)
Sustainably harvested
Takeback
Thermal bridging
Thermal mass
Urbanization
Urea formaldehyde
Vapor-resistant
Virgin materials
Volatile organic compounds (VOCs)
Walk-off mats
Waste separation
Water efficiency
Water footprint
Water-based
Water-resistant
Waterproof
Wetlands
Xeriscaping
Zoning

Industry Recognized Credentials

If you are training through an NCCER-accredited sponsor, you may be eligible for credentials from NCCER's Registry. The ID number for this module is 70101-15. Note that this module may have been used in other NCCER curricula and may apply to other level completions. Contact NCCER's Registry at 888.622.3720 or go to **www.nccer.org** for more information.

Contents

Topics to be presented in this module include:

Contents (continued)

Figures

Figures (continued)

1.0.0 IMPROVING YOUR PERSONAL ENVIRONMENTAL IMPACT

Objective

Select actions to improve your personal environmental impact at home and work.

  a. Describe the major challenges buildings cause directly or indirectly on the green environment.
  b. Identify choices in your personal and work life that impact the green environment.
  c. Prioritize your actions in terms of which ones matter most for the green environment.

Trade Terms

Acidification: A process that converts air pollution into acid substances, leading to acid rain. Acid rain is best known for the damage it causes to forests and lakes. Also refers to the outflow of acidic water from metal and coal mines.

Aquifer: An underground layer of water-bearing rock or soil from which groundwater can be extracted using a well.

Aquifer depletion: A situation where water is withdrawn from its underground source faster than the rate of natural recharge.

Biodiversity: A measure of the variety among organisms present in different ecosystems.

Carbon cycle: The movement of carbon between the biosphere, atmosphere, oceans, and geosphere of the Earth. It is a biogeochemical cycle. In the cycle, there are sinks, or stores, of carbon. There are also processes by which the various sinks exchange carbon.

Carbon footprint: A measure of impact human activities have on the environment. It is determined by the amount of greenhouse gases produced. It is measured in pounds or kilograms of carbon dioxide.

Carbon neutral: An action or product whose production absorbs as much carbon from the atmosphere as it produces.

Carbon offset: An agreement with another party that they will reduce their carbon production by some amount in exchange for payment.

Carpool: An arrangement in which several people travel together in one vehicle. The people take turns driving and share in the cost.

Compact fluorescent lamp (CFL): A fluorescent bulb that is designed to fit in a normal light fixture. They use less energy and last longer than incandescent bulbs.

Conservation: Using natural resources wisely and at a slower rate than normal.

Deforestation: The removal of trees without sufficient replanting.

Desertification: The creation of deserts through degradation of productive land in dry climates by human activities.

Ecological footprint: A measure of impact human activities have on the environment. It compares human consumption of natural resources with the Earth's capacity to regenerate them. Measured in hectares or acres.

Ecosystem: A combination of all plants, animals and microorganisms in an area that complement each other. These function together with all of the nonliving physical factors of the environment.

Embodied energy: The total energy required to bring a product to market. It includes raw material extraction, manufacturing, final transport, and installation.

Energy efficiency: Getting more use out of electricity already generated.

ENERGY STAR: A United States government program to promote energy efficiency. It is a joint program of the US Environmental Protection Agency (EPA) and the US Department of Energy (DOE).

Equipment idling: The operation of equipment while it is not in motion or performing work. Limiting idle times reduces air pollution and greenhouse gas emissions.

Erosion: The displacement of solids by wind, water, ice, or gravity or by living organisms. These solids include rocks and soil particles.

Geothermal: Heat that comes from within the Earth.

Global climate change: Changes in weather patterns and temperatures on a planetary scale. This may lead to a rise in sea levels, melting of polar ice caps, increased droughts, and other weather effects.

Greenhouse effect: The overall warming of a planet's surface due to retention of solar heat by its atmosphere.

Hybrid vehicle: A vehicle that uses two or more power sources for propulsion.

Life cycle: The useful life of a system, product, or building.

Life cycle assessment: An analytic technique to evaluate the environmental impact of a system, product, or building throughout its life cycle. This includes the extraction or harvesting of raw materials through processing, manufacture, installation, use, and ultimate disposal or recycling.

Life cycle cost: The cost of a system or a component over its entire life span.

Light-emitting diode (LED) lamp: A highly efficient, electronic light bulb using a glowing diode designed to fit a standard light fixture.

Low-emission vehicle: Vehicles that produce fewer emissions than the average vehicle. Beginning in 2001, all light vehicles sold nationally were required to meet this standard.

Ozone depletion: A slow, steady decline in the total amount of ozone in Earth's stratosphere.

Ozone hole: A large, seasonal decrease in stratospheric ozone over Earth's polar regions. The ozone hole does not go all the way through the layer.

Payback period: The amount of time it takes to break even on an investment; the period of time after which savings from an investment equals its initial cost.

Phantom loads: Electricity consumed by appliances and devices when they are switched off.

Photovoltaic: A technology that converts light directly into electricity.

Preservation: The act and advocacy of protecting the natural environment.

Raw material: A material that has been extracted directly from nature. It is in an unprocessed or minimally processed state.

Recycling: The reprocessing of old materials into new products. A goal is to prevent the waste of potentially useful materials and reduce the consumption of new materials.

Renewable: A resource that may be naturally replenished.

Solar: Energy from the sun in the form of heat and light.

Sprawl: Unplanned and inefficient development of open land.

Stormwater runoff: Unfiltered water that reaches streams, lakes, and oceans after a rainstorm by flowing across impervious surfaces.

Urbanization: Converting rural land to higher density development.

Water efficiency: Managing water use to prevent waste or overuse. Includes using less water to achieve the same benefits.

Water footprint: The volume of water used directly or indirectly to sustain something over a period of time.

Our impacts on the green environment are considerable. Resource use and global climate change are major concerns. Global climate change, also called global warming, involves changes in weather and temperature worldwide. Your daily activities have an impact on these larger problems, with your carbon footprint being a measure of your contribution to global climate change. The products you buy and the energy you consume affect the climate. It is also influenced by waste you throw away and gasoline you use. Many opportunities exist to reduce your carbon footprint, which include reducing energy and fuel use and rejecting, reducing, reusing, and recycling products. You can also plant vegetation or find better energy sources.

1.1.0 The Nature of Change

The man-made environment is changing along with the green environment. Some changes are independent of human actions. The changes in the seasons, for example, are a result of the Earth's position in relation to the sun. These changes have predictable effects on temperature and weather around the world. *Figure 1* shows colored leaves that mark the change of summer to autumn in northern climates. Seasonal changes make the setting for each building unique. Development in each region should respond to

70101-14_F01.EPS

Figure 1 Colored leaves mark the changing of seasons in many regions.

local climate, and should also reflect the demands of that climate on the built environment.

Other changes are less predictable, such as severe weather patterns. In the United States, hurricanes are common in summer and fall; snow and ice storms occur in the winter, and tornados are more common in spring and summer. In other places, cyclones or typhoons are a threat. Flooding and volcanic eruptions cause extensive damage when they occur. *Figure 2* shows a satellite view of a hurricane, also known as a cyclone or typhoon. This is one of the most severe types of weather affecting life on earth.

Severe weather also causes a challenge for the built environment. Buildings must be able to function through severe weather. The role of buildings is to preserve the safety and security of their occupants, making facilities that can survive severe storms important.

1.1.1 Changes in the Man-Made Environment

Just as the green environment changes, the built environment also changes. The role of the built environment is to help humans survive and prosper, as well as to provide shelter from the effects of weather and climate. Over time, buildings have become larger and more complex in order to meet changing human needs and desires. New technologies and practices make buildings more functional, comfortable, and efficient, while other developments result in new styles and appearances.

Expectations for built facilities have also changed. According to the National Association of Home Builders, the average house size has almost doubled in the past forty years. Houses have grown from an average of 1,400 square feet (130 square meters) in 1970 to nearly 2,700 square feet (250 square meters) in 2013. During that time, the average family size in the United States decreased from 3.14 people per household to 2.54. This means that the amount of living space per person has risen nearly 50 percent over the past 25 years.

At the same time, the development of buildings and neighborhoods has had an impact on human health and well-being. The Federal Highway Administration has found that Americans walk for fewer than 6 percent of daily trips. Much of the development in the United States is designed around the automobile. The absence of sidewalks and bicycle paths in neighborhoods discourages people from walking. Large distances between residential areas and services such as stores and schools make people more likely to drive. A University of Maryland study investigated the health effects of urban sprawl. It found that people who live in neighborhoods characterized by urban sprawl weigh about six pounds more than people living in compact neighborhoods. (Compact neighborhoods have sidewalks and shops close to residential areas.) More recent studies have confirmed the relationship between our location and our personal health and well-being. Location also has implications on the green environment. For example, many developments have been built in locations without enough water to support growth (*Figure 3*).

1.1.2 Relationships Between Human Activities and the Green Environment

Human population is growing and expectations for standards of living are increasing. This creates a greater demand for resources from the green environment. The raw materials used to build, furnish, and operate buildings are extracted

70101-14_F03.EPS

Figure 3 Las Vegas has been developed in a location without much water. Its sprawl leads to increased transportation requirements.

70101-14_F02.EPS

Figure 2 Satellite view of a hurricane.

Worldwide Power Use is Growing at a Staggering Rate

According to the US Energy Information Administration, the world's total power use is expected to increase by 56 percent between 2010 and 2040. Over 85 percent of that growth will come from projects in rapidly developing countries, such as Colombia (shown here).

70101-14_SA01.EPS

from the green environment. The fuel required to power cars, factories, and buildings is also extracted from the environment, while burning these fuels has consequences as well.

The impacts of human activities on the green environment are significant; you can see some impacts in your own backyard. There are also global impacts which have effects that may not be seen for many years (*Figure 4*). Either way, the impacts of human beings on the green environment are undeniable. Major environmental challenges include the following:

- *Global climate change* – Increasing greenhouse gases produce an overall rise in global temperatures. These gases are due in part to the burning of fossil fuels as a result of urbanization. Urbanization often increases air pollution. Temperature changes affect sea levels as ice melts at the Earth's poles, and also increase the potential for severe weather.
- *Excess wood harvesting* – The harvesting of wood resources at an unsustainable rate is known as deforestation. It leads to soil depletion, pollution of streams, and habitat loss. Deforestation also contributes to global climate change because trees reduce greenhouse gases. Preservation of forest areas can help to slow global warming.
- *Species extinction* – Habitat loss has caused thousands of species to become extinct. This is known as a loss of biodiversity. The remaining habitats are fragmented and degraded in quality. Human beings depend on diverse ecosystems to purify the air and water. These ecosystems help to stabilize climate change and provide a variety of resources from lumber to medicine. Ecosystem health is essential for human survival.
- *Decreasing water supplies* – The rise in global temperatures has caused desertification (the spread of deserts) in dryer parts of the world. In other areas, overgrazing and the overuse of groundwater has led to aquifer depletion. This is particularly true in agricultural and urban areas. Aquifers are the only reliable source of water in many parts of the country.
- *Air and water pollution* – Industrial activities and the daily activities of people contribute to the pollution of air and water. The burning of fossil fuels for transportation and energy

(A) REFINING   (B) MINING   (C) GRADING

(D) DESERTIFICATION   (E) CLEAR CUTTING   (F) LOGGING

70101-14_F04.EPS

Figure 4 The impacts of human activities on the green environment.

produces smog, which leads to **acidification** (acid rain) and plant decline. The use of these fuels also contributes to ground level ozone and other forms of air pollution. Runoff from paved and deforested areas contributes to water pollution. Industrial deposits into waterways increase water temperature and contaminants. Many modern components of wastewater cannot be removed using normal treatment methods. These include such things as prescription drugs and plastic residuals, which accumulate in nature with unknown long-term consequences.

- *Soil contamination and depletion* – Contaminants from human activities often remain in the soil and eventually migrate to water sources. Sources of contamination range from leaking underground storage tanks to **stormwater runoff** from paved areas. Removing vegetation exposes topsoil and causes **erosion** losses.
- *Loss of ozone* – Release of chlorine-based gases such as refrigerants has led to **ozone depletion** in the upper atmosphere. Ozone molecules are necessary to shield living organisms from **solar** radiation. Reduced levels of ozone are thought to be responsible for the increased

rate in skin cancer and cataracts in humans as well as damage to marine and terrestrial eco-systems. These chemicals also contribute to a greater-than-usual growth in the seasonal **ozone hole** that appears naturally in the polar regions.

None of these occur in isolation. In many cases, human activities contribute to more than one problem at a time. All of them require attention to ensure the health of the planet for future generations.

1.2.0 Impact of Individual Human Activities

Given the growing awareness of human impacts on the green environment, how do your daily activities contribute to environmental problems and their solutions? The manufacturing, transportation, use, and disposal of the products used in daily living contribute to each person's individual impacts. Your individual carbon footprint can serve as a measure of how your lifestyle contributes to global climate change. Your work on the job site and the buildings you help to create also contribute to the challenges being faced by the green environment.

1.2.1 The Average American Household

The average US household is annually responsible for the production of 4,060 pounds (1,842 kilograms) of garbage, over 46,000 gallons (174,129 liters) of wastewater, and over 96,000 pounds (43,545 kilograms) of carbon dioxide (CO_2), along with smaller amounts of sulfur dioxide (SO_2), nitrogen oxides (NO_X), and heavy metals. These impacts come from the use of products, the use of resources such as electricity, the types of waste generated, and travel (*Figure 5*).

The Consumer Electronics Association estimates that the average household spends nearly $1,200 per year on electronics, including televisions, digital cameras, and other devices. According to the US Energy Information Administration (EIA), the average household also spends about $1,400 per year to run these devices, along with all the other uses of energy in the home such as heating and lighting. In fact, according to studies conducted by Nielson Media Research, the average American household has more televisions than it does people. The US Department of Agriculture reports that American households also spend an average of more than $6,500 per year on food, much of which requires extensive use of water and fossil fuel energy to produce, process, and transport to the table.

According to the EPA, the average household in the United States spends around $520 per year on its water and sewer bill. This varies significantly based on cost of water in different parts of the country. Each person in the United States consumes 100 gallons (379 liters) per day on average and generates over 50 gallons (189 liters) of wastewater. (About half of the water consumed is for irrigation and goes back into the ground rather than the wastewater stream.) Of this wastewater, 9,000 gallons (34,069 liters) per year is used to flush away only 230 gallons (871 liters) of waste from toilets, a surprisingly wasteful use of water that has been treated to drinking water standards. Water use also requires energy for pumping, collection, treatment, and distribution. For example, letting a faucet run for five minutes uses as much energy as leaving a 60-watt light bulb on for 14 hours.

People also produce tremendous amounts of solid waste, both directly in homes and indirectly through the production of consumer goods. The average American generates 4.38 pounds (2 kilograms) of garbage per day, of which 75 percent could be recycled, but less than 35 percent typically is. That adds up to nearly 1,600 pounds (726 kilograms) per person per year. On average, each American consumes 3 gallons (11.4 liters) of gasoline per day—enough to fill up 21 bathtubs per year. This doesn't count all the trips on public transit, trains, buses, or airplanes.

All this adds up to a considerable impact on the green environment. To determine exactly how

70101-14_F05.EPS

Figure 5 Personal choices make a difference.

much impact you and your family have, complete the inventory of household impacts found in *Worksheet 1A* and *B*.

1.2.2 The Impacts of the Products You Use

Every product you use has a hidden history of harvesting, extraction, manufacturing, and transportation that cause impacts beyond what you see in the product itself. The total energy required to bring a product to market is its **embodied energy**. It includes everything from raw material extraction to manufacturing to final transport and installation. For example, aluminum beverage cans may be manufactured with metals mined in several different countries and then shipped on pallets made from wood harvested in another country, using fuel from yet another country. The can itself is more costly and complicated to manufacture than the beverage. Drinking the beverage takes a few minutes, throwing the can away takes a second, and yet the true impact is immense. Think of all the products you consume or use over the course of a given year. How much of an impact do you think those products have? Answer the questions in *Worksheet 2* to develop an

---

WORKSHEET 1A: INVENTORY YOUR HOUSEHOLD IMPACTS

Conducting an inventory of your household consumption, waste generation, and activities is the first step in understanding how you can reduce your impact. Answer the following questions based on your best guess. If you share a household with other people, divide the total answer for your household by the number of people who live in your home.

How many gallons of garbage do you throw away each week? An average American garbage can holds 32 gallons. Multiply by 52 to calculate the gallons of garbage you throw away per year.

Gallons of garbage per year: _____

How much electricity do you use per year? You'll find this information on your electricity bill. You can add up the total for 12 months worth of bills, or multiply a monthly average by 12. If you know how much your electricity costs per month, you can estimate the amount of energy used in kilowatt-hours by dividing your total bill by 10.

Total electricity per year in kilowatt-hours: _____

How many therms of natural gas do you use per year? Check your natural gas bill if you get one. It will tell you how many therms of gas you use per month. Your highest values will probably be during the winter heating season. If you know your monthly average, multiply by 12 to calculate your annual use.

Total therms of natural gas per year: _____

How many gallons of propane do you use each year? You can check this on your propane bill if you get one. Add up the total number of gallons per year.

Total gallons of propane per year: _____

How many gallons of fuel oil do you use per year? You can check this on your fuel oil bill if you get one. Add up the total number of gallons per year.

Total gallons of fuel oil per year: _____

On average, what is your monthly combined water and sewage bill? Check your monthly bill if you get one and add up all the amounts for a one-year period. If you don't get a water and sewage bill, you can estimate this amount as approximately $75 per person per year.

Annual cost for water and sewage: _____

How many square feet is your house or dwelling? If you don't know, draw a floor plan of your house or apartment and estimate the floor area in square feet.

Size of house (floor area) in square feet: _____

70101-14_WS01A.EPS

On average, how many miles do you drive your household vehicles per week, and what are their average fuel efficiencies? 300 miles per week per vehicle or 15,000 miles per year is about average in the United States. If you're not sure about fuel efficiency, assume your car gets 22 miles to the gallon, which is about average. If you know how many gallons of fuel you use per week, skip directly to the end of the line.

Car 1 miles per week _____ / *Car 1 miles per gallon* _____ = *Car 1 gallons per week* _____

Car 2 miles per week _____ / *Car 2 miles per gallon* _____ = *Car 2 gallons per week* _____

Car 3 miles per week _____ / *Car 3 miles per gallon* _____ = *Car 3 gallons per week* _____

Add up the gallons of gas per week for all your vehicles, and then multiply by 52 to estimate the gallons of gasoline you use per year.

Total gallons of gasoline you use per year: _____

On average, how much do you travel each year on airplanes? Estimate the number of flight segments below for each of the three distances. A round trip counts as two segments, and each segment of a multi-segment flight counts as its own flight. For example, if you fly from Baltimore to Cincinnati to Atlanta, that counts as two flight segments.

Number of short-haul flight segments (less than 700 miles or 2 hours): _____

Number of medium-haul flight segments (700 – 2,500 miles or 2 – 4 hours): _____

Number of long-haul flight segments (more than 2,500 miles or longer than 4 hours): _____

On average, how many miles do you travel on public transportation per year?

Number of miles per year on transit bus/subway: _____

Number of miles per year on intercity bus: _____

Number of miles per year on intercity train: _____

Add up the total number of miles per year on public transportation: _____

Your answers to these questions will help you calculate your own carbon footprint later in the module. Keep track of your answers in the workbook or on a separate worksheet.

70101-14_WS01B.EPS

inventory of the products you consume and begin to estimate the impacts those products have. Your answers to these questions will help you calculate your carbon footprint later in this module.

1.2.3 Your Carbon Footprint and Global Climate Change

Greenhouse gases such as CO_2 create what is known as the greenhouse effect (*Figure 6*). A certain level of greenhouse gas is essential because it prevents the loss of heat into outer space. However, if the level increases too much, it can result in global warming. The carbon cycle is the transfer of carbon (mostly in the form of CO_2) between Earth and the atmosphere. Plants absorb CO_2 and release oxygen, while people (and many of their energy-using activities, such as combustion) and animals absorb oxygen and release CO_2. The delicate balance of this cycle has shifted as people generate more and more CO_2 while depleting trees and plants that absorb CO_2 and generate oxygen.

The inventories you created are useful for estimating your own contribution to the greenhouse effect. Any changes made now can help to reduce the severity of these impacts.

For example, the rise in global temperatures is expected to cause a certain portion of polar ice caps to melt. This will cause a rise in sea levels as water from melting ice runs off into oceans. If you live in a coastal area at a low elevation, you might find that the land around you is disappearing as the ocean level rises (*Figure 7*). At a minimum, you can expect severe weather to occur more frequently—watch for stronger, more frequent

WORKSHEET 2: INVENTORY YOUR PRODUCT IMPACTS

Consider the products you buy or that are bought for you over the course of a year. Answer the following questions based on your best guess. If you share a household with other people, divide the total answer for your household by the number of people who live in your home.

Eating out: $ _____ *per month* × 12 = $ _____ *per year*

Meat, fish, & protein: $ _____ *per month* × 12 = $ _____ *per year*

Cereals & baked goods: $ _____ *per month* × 12 = $ _____ *per year*

Dairy: $ _____ *per month* × 12 = $ _____ *per year*

Fruits & vegetables: $ _____ *per month* × 12 = $ _____ *per year*

Other: $ _____ *per month* × 12 = $ _____ *per year*

On average each month, how much do you spend on the following goods and services? Multiply each amount by 12 to estimate your average annual spending in each category. If you already know how much you spend per year, you can skip the monthly amount and write the average amount per year at the end of each line.

Clothing: $ _____ *per month* × 12 = $ _____ *per year*

Furnishings & household items: $ _____ *per month* × 12 = $ _____ *per year*

Other goods: $ _____ *per month* × 12 = $ _____ *per year*

Services: $ _____ *per month* × 12 = $ _____ *per year*

70101-14_WS02.EPS

hurricanes, tornados, and storms interspersed with periods of drought. There are also many other potential effects of global climate change that scientists believe might occur as temperatures rise. These include increases in forest fires due to drier conditions and possible extinction of species such as polar bears due to a loss in habitat. Many scientists believe that increases in greenhouse gases will lead to rising temperatures worldwide. Others believe that climate change may have drastically different effects in different parts of the world—some areas may get much warmer, while others may get much colder. One scientist even refers to the concept as "global weirding" since it is difficult to predict exactly how global systems will respond. While the ultimate effects of increased greenhouse gases are yet to come, scientists agree that human activities are contributing to climate change. One way to evaluate your personal impact on global climate change is through

Did You Know?

Aluminum Cans

The United States still gets three-fifths of its aluminum from virgin ore, at twenty times the energy intensity of recycled aluminum. Americans throw away enough aluminum to replace our entire commercial aircraft fleet every three months.

70101-14_SA02.EPS

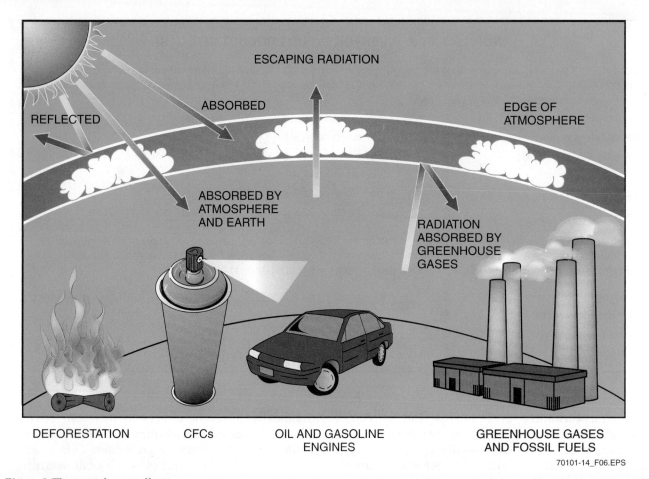

ESCAPING RADIATION

ABSORBED

REFLECTED

EDGE OF
ATMOSPHERE

ABSORBED BY
ATMOSPHERE
AND EARTH

RADIATION
ABSORBED BY
GREENHOUSE
GASES

DEFORESTATION   CFCs   OIL AND GASOLINE   GREENHOUSE GASES
                                      ENGINES   AND FOSSIL FUELS

70101-14_F06.EPS

Figure 6 The greenhouse effect.

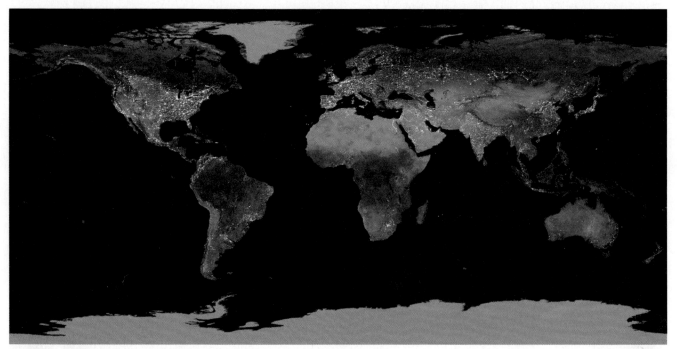

70101-14_F07.EPS

Figure 7 The effects of rising sea levels.

footprint analysis. There are three different types of footprint analysis: carbon footprint, ecological footprint, and water efficiency/water footprint. Your carbon footprint evaluates how many tons of carbon are emitted as a result of your actions. Your ecological footprint is measured in acres or hectares and represents the amount of productive land area it takes to support you, including manufacturing the products you buy, growing the food you eat, and absorbing the waste you produce. Your water footprint estimates the number of gallons or liters of water per year required to sustain your lifestyle. Each of these methods focuses on a different way to calculate the effects you personally have on the green environment. Your carbon footprint is a measure of your contribution to global climate change. Your ecological and water footprints focus on how much of the Earth's limited resources you consume. In fact, your ecological footprint is often expressed in terms of how many Earths it would take if everyone lived like you. The average American's ecological footprint is about 20 acres (8 hectares), compared to a world average of 6.7 acres (2.7 hectares).

Unfortunately, if only 12 percent of the Earth's biosphere or viable surface area is reserved for other species, there are less than 5 acres (2 hectares) available per person. With continued population growth and increase in standards of living, the amount of overshoot stands to increase even further. Right now, 20 percent of the Earth's viable ecological footprint is consumed by 2.5 percent of its population—the richest 2.5 percent. Many of the remaining population do not have enough for basic survival. Some cities are making concerted efforts to reduce their ecological footprints (*Figure 8*).

Using the information collected earlier in this module about household activities and consumption, you can estimate how many pounds of carbon per year are generated by each of these activities. For each of the sections in *Worksheet 3*, fill in the amounts from your earlier inventories and multiply by the factors shown to estimate the total pounds of carbon for each activity. Then add up the totals for each category to estimate your total carbon footprint in pounds.

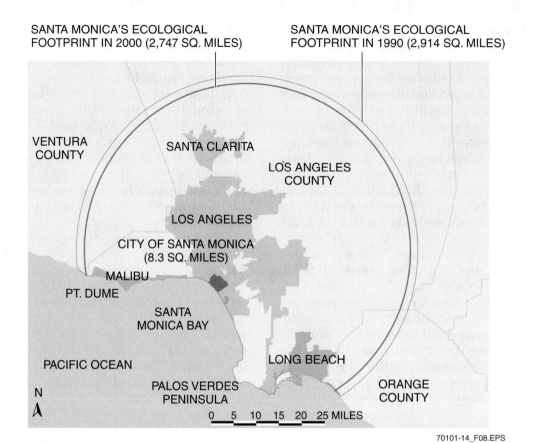

SANTA MONICA'S ECOLOGICAL
FOOTPRINT IN 2000 (2,747 SQ. MILES)

SANTA MONICA'S ECOLOGICAL
FOOTPRINT IN 1990 (2,914 SQ. MILES)

VENTURA
COUNTY

SANTA CLARITA

LOS ANGELES
COUNTY

LOS ANGELES

CITY OF SANTA MONICA
(8.3 SQ. MILES)

MALIBU

PT. DUME

SANTA
MONICA BAY

PACIFIC OCEAN

N

LONG BEACH

PALOS VERDES
PENINSULA

ORANGE
COUNTY

0   5   10   15   20   25 MILES

70101-14_F08.EPS

Figure 8 Santa Monica reduced its ecological footprint through recycling and energy conservation programs.

According to the World Bank, in 2010 the average American person generated about 38,800 pounds (17.6 metric tons) per person, compared to 20,060 pounds (9.1 metric tons) by the average German, 8,400 pounds (3.8 metric tons) by the average Mexican, and 660 pounds (0.3 metric tons) by the average Kenyan.

1.3.0 Things You Can Do to Make a Difference

You can do many things to reduce your personal impact on the green environment. Each action has a different level of impact. For example, according to the EPA, planting a tree seedling and allowing it to grow for 10 years can absorb about 86 pounds (0.04 metric tons) of CO_2. Avoiding the use of one gallon of gasoline saves about 19 pounds (0.009 metric tons) of CO_2, while taking an average car off the road for one year saves about 11,765 pounds (5.3 metric tons) of CO_2. You may not be able to stop driving, but you can probably stop driving as much. Small changes such as these are known as leverage points.

1.3.1 Seeking Leverage Points

Leverage points are small changes that make a big difference. To understand leverage points, think about changing a tire on your car (*Figure 9*). Placing a jack in the right location under the frame enables you to lift a heavy car off the ground. The jack and its placement are the critical factors that allow you to lift an enormous load.

GOING GREEN

Footprint Calculators

To calculate your ecological footprint, go to **www.myfootprint.org**. An online carbon footprint calculator is available at **www.carbonfootprint.com**. You can calculate your water footprint at **www.waterfootprint.org**. To see how much your own footprint equals in terms of impacts, go to **www.epa.gov/cleanenergy** and use the Greenhouse Gas Equivalencies Calculator to convert your footprint to equivalents such as tanker trucks' worth of gasoline or railcars full of coal.

WORKSHEET 3: DETERMINE YOUR CARBON FOOTPRINT

Fill in the amounts from Worksheets 1 and 2 to calculate your total carbon footprint in pounds of carbon. Which of your activities contributes the most to your carbon footprint?

Item	Quantity	Carbon Factor	Total Pounds of CO_2/Year
Gallons of garbage/year:	_____	× 2 lbs/gallon	= _____ lbs/year
Total electricity/year in kilowatt-hours:	_____	× 1.4 lbs/kWh	= _____ lbs/year
Total therms of natural gas/year:	_____	× 11.7 lbs/gallon	= _____ lbs/year
Total gallons of propane/year:	_____	× 12.7 lbs/gallon	= _____ lbs/year
Total gallons of fuel oil/year:	_____	× 22.4 lbs/gallon	= _____ lbs/year
Annual cost for water and sewage:	_____	× 8.9 lbs/dollar	= _____ lbs/year
House size (floor area) in square feet:	_____	× 2.1 lbs/sq ft	= _____ lbs/year
Gallons of gasoline/year:	_____	× 20 lbs/gallon	= _____ lbs/year
No. of short-haul flight segments/year:	_____	× 304 lbs/segment	= _____ lbs/year
No. of medium-haul flight segments/year:	_____	× 726 lbs/segment	= _____ lbs/year
No. of long-haul flight segments/year:	_____	× 2,217 lbs/segment	= _____ lbs/year
Miles per year on public transportation:	_____	× 0.5 lb/mile	= _____ lbs/year
Eating out (US dollars/year):	_____	× 0.8 lb/dollar	= _____ lbs/year
Meat, fish, & protein (US dollars/year):	_____	× 3.2 lbs/dollar	= _____ lbs/year
Cereals & baked goods (US dollars/year):	_____	× 1.6 lbs/dollar	= _____ lbs/year
Dairy (US dollars/year):	_____	× 4.2 lbs/dollar	= _____ lbs/year
Fruits & vegetables (US dollars/year):	_____	× 2.6 lbs/dollar	= _____ lbs/year
Other (US dollars/year):	_____	× 1.0 lb/dollar	= _____ lbs/year
Clothing (US dollars/year):	_____	× 1.0 lb/dollar	= _____ lbs/year
Household items (US dollars/year):	_____	× 1.0 lb/dollar	= _____ lbs/year
Other goods (US dollars/year):	_____	× 0.75 lb/dollar	= _____ lbs/year
Services (US dollars/year):	_____	× 0.4 lb/dollar	= _____ lbs/year
Total Carbon Footprint		**=**	**_____ lbs/year**

70101-14_WS03.EPS

When deciding what changes to make, consider how much effort is required to make the change, how easy it will be to sustain, and how much impact the change will have. Look for simple changes that you can make (such as using a jack) and the best opportunities to make them that will have the most impact (such as selecting the right place on the car to position the jack). Changes that require you to adjust how you behave on an ongoing basis are often more difficult than changes requiring a single effort.

Finally, be sure to consider all the associated costs and benefits of potential actions, not just the initial costs of purchase. Sometimes the life cycle costs of an item make it worthwhile to spend more money up front to achieve greater savings over time. Life cycle assessment involves looking at the environmental impact throughout a product's entire life cycle. For example, an energy-efficient light bulb may cost more initially but will pay for itself through reduced energy use. It may also save money through reduced maintenance costs because it lasts longer and therefore won't have to be replaced as often, and it may reduce air conditioning costs because it runs cooler. The amount of time required to save the additional money invested up front is known as the payback period.

70101-14_F09.EPS

Figure 9 Leverage points are like a carefully placed car jack.

1.3.2 Reducing Energy Use

One important way to reduce your carbon footprint is by reducing your overall energy use. About 87 percent of all energy used in the United States comes from burning fossil fuels, including oil, coal, natural gas, and propane (*Figure 10*). In fact, the burning of fossil fuels is the largest single contributor to carbon footprint in the United States, and it is at the root of many of the challenges faced by the green environment.

The average US household spends its energy dollar as shown in *Figure 11*. It is not surprising that most of the money goes toward heating, cooling, hot water, appliances, and lighting. What may be surprising is the large *Other* category. This category includes the growing number of battery chargers used to power cellular phones, portable tools, and other electronic devices. These chargers draw a small amount of power even when the battery is fully charged. The warmth you feel when you touch a battery charger is some of the wasted power being dissipated as heat.

Look around—how many items plugged into your walls have a clock, timer, or charger built in? All of these items draw what are called phantom loads. In other words, they are consuming energy even when it appears that the device itself is turned off. The only way to stop the power consumption by these devices is to unplug them completely. There are many ways to reduce your use of energy. Some of the methods recommended by **energy.gov** are listed below.

To reduce the amount of energy used to heat and cool your home, do the following:

- Choose energy-efficient furnaces or air conditioners that are the right size for your home. Check for rebates from your local utility that can help pay to replace your current systems with more efficient ones.

GOING GREEN

Reducing your Personal Carbon Footprint

Even simple changes can make a big difference in your carbon footprint. Here are some ideas from the US Environmental Protection Agency on how much carbon you can save. To convert to metric tons, divide the number of pounds by 0.000453.

By...	You can save...
Reducing the number of car trips you take	One pound of CO_2 for every mile you don't drive
Adjusting your thermostat two degrees warmer in summer and two degrees cooler in winter	2,000 pounds of CO_2 on average per year
Unplugging chargers and turning off appliances and computers when not in use	1,000 pounds of CO_2 per year
Shortening your shower by two minutes per day	342 pounds of CO_2 per year
Updating your refrigerator to a new ENERGY STAR model	500 pounds of CO_2 per year
Using a reusable mug instead of disposable cups	135 pounds of CO_2 per year if done twice per day, every day
Recycling solid waste	One pound of CO_2 for every pound of waste recycled

U.S. ENERGY CONSUMPTION BY SOURCE

 BIOMASS RENEWABLE 5.7%
HEATING, ELECTRICITY, TRANSPORTATION

 PETROLEUM NONRENEWABLE 19.3%
TRANSPORTATION, MANUFACTURING

 HYDROPOWER RENEWABLE 3.1%
ELECTRICITY

 NATURAL GAS NONRENEWABLE 30.5%
HEATING, MANUFACTURING, ELECTRICITY

 GEOTHERMAL RENEWABLE 0.3%
HEATING, ELECTRICITY

 COAL NONRENEWABLE 24.5%
ELECTRICITY, MANUFACTURING

 WIND RENEWABLE 2.0%
ELECTRICITY

 URANIUM NONRENEWABLE 10.1%
ELECTRICITY

 SOLAR & OTHER RENEWABLE 0.4%
LIGHT, HEATING, ELECTRICITY

PROPANE NONRENEWABLE 4.2%
MANUFACTURING, HEATING

70101-14_F10.EPS

Figure 10 US energy consumption by source (US Department of Energy).

- Properly insulate your home's attic, walls, and slab or crawl space, including all ducts outside the conditioned space of your home.
- Contact your utility company for a free energy audit, or go online to **http://hes.lbl.gov** for a web-based energy audit tool that can help identify areas where you can improve household energy use.

- Install programmable thermostats, insulated windows, and ceiling fans to increase comfort in your home. Programmable thermostats can save up to 20 percent of your heating and cooling costs by cutting back your systems at night and during the day when you are not at home.

Think About It

Living Off the Grid

Suppose you want to build a small house in the country. The house will provide a retreat and will have all the comforts of home, but with its rural location, it is too far from existing power supplies to be connected to the grid. How will you provide power for your house and its amenities? Which electrical devices are you willing to live without in order to save energy?

70101-14_SA04.EPS

ENERGY CONSUMED (KWH/YEAR)

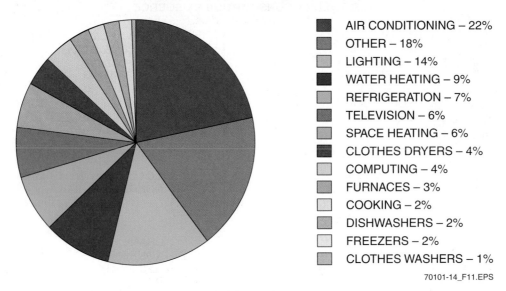

- AIR CONDITIONING – 22%
- OTHER – 18%
- LIGHTING – 14%
- WATER HEATING – 9%
- REFRIGERATION – 7%
- TELEVISION – 6%
- SPACE HEATING – 6%
- CLOTHES DRYERS – 4%
- COMPUTING – 4%
- FURNACES – 3%
- COOKING – 2%
- DISHWASHERS – 2%
- FREEZERS – 2%
- CLOTHES WASHERS – 1%

70101-14_F11.EPS

Figure 11 Energy consumption in the average US household.

To save energy used for hot water heating, do the following:

- Consider replacing your tank hot water heater with a tankless or on-demand hot water heater that only heats water as needed, or a solar hot water heater if your home design can accommodate one.
- Set your water heater to 120°F (49°C) instead of 140°F (60°C). Most appliances that require hotter water, such as dishwashers, have built-in heating coils.
- Insulate your hot water pipes and hot water tank to reduce heat loss.
- Replace faucets and showerheads with low-flow models if they are more than 10 years old. Newer showerheads can deliver a pleasant shower at 1.5 gallons (5.7 liters) per minute or less due to the introduction of air into the water stream.

To save energy used by appliances such as washers, dryers, refrigerators, and dishwashers, do the following:

- Wash your laundry in cold water. It helps your clothes last longer and fade less, and can save over $60 per year in hot water heating costs.
- Consider a front-loading washing machine. These machines save up to 25 percent of the water used for washing, along with the energy required to heat that water. They also use less detergent and remove more water from clothes during the spin cycle, reducing drying time and energy use.
- Use a clothesline instead of a dryer.
- Choose a top-freezer model of refrigerator instead of a side-by-side model. Top-freezer

models are more efficient due to the placement of the compressor. Beware of built-in icemakers—they add to the energy cost and are the culprit for many repair calls.

- Keep your appliances full. Full refrigerators maintain more even temperatures and cycle on less frequently. Full washing machines and dishwashers get the maximum amount of cleaning possible for the least use of resources. Load washing machines, dryers, and dishwashers according to the manufacturer's instructions to increase effectiveness.

To save energy in lighting, household electronics, and miscellaneous areas, do the following:

- Swap your incandescent bulbs for compact fluorescent lamps (CFLs) or light-emitting diodes (LEDs). They use less energy and last longer than a standard incandescent bulb.
- Use power strips and sensors. Power strips can be used to unplug your electronic devices without disrupting cords and cables. Electric photocells and occupancy sensors turn lights off automatically when they are not needed.
- Buy ENERGY STAR appliances and devices. These products perform in the top 25 percent of their product class in terms of energy use and contribute to overall energy conservation (*Figure 12*). Listings of approved products are available online at **www.energystar.gov**.
- Use the sleep mode or shut down equipment such as computers when not in use. Keeping a computer and monitor running 24 hours a day consumes up to 1,100 kilowatt-hours of energy, which could cost over $100 per year depending on electricity rates.

Think About It

Life Cycle Costs and Benefits

Consider the light bulbs shown here, all equivalent to a 60-watt incandescent bulb. The 8.5 watt light-emitting diode (LED) lamp on the right costs about $10, lasts for 25,000 hours of use or more, and consumes about $24 worth of energy over its life cycle. The 60-watt incandescent lamp on the left costs about 50 cents and lasts for about 2,000 hours of use. Over 25,000 hours of use (requiring 13 bulbs), this bulb will consume about $170 worth of energy at 11.3 cents per kilowatt-hour (typical in the United States). The 15-watt compact fluorescent lamp in the center costs

70101-14_SA05.EPS

about $1.70, lasts for about 10,000 hours of use, and would consume about $42 worth of energy over a life cycle of 25,000 hours (not including the costs of three bulbs needed to last this long). All generate about the same amount of light. Which one is the best investment?

1.3.3 Reducing Fuel Use and Increasing Fuel Efficiency

Over half of the carbon footprint of most Americans is due to transportation choices. Some ways to reduce the overall use of fuel and increase fuel efficiency include the following:

- Drive a fuel-efficient, low-emission vehicle. Many options are available, including gasoline-electric hybrid vehicles that get over 50 miles per gallon (21.2 kilometers per liter). You may also consider driving a scooter or motorcycle.

70101-14_F12.EPS

Figure 12 ENERGY STAR products meet strict requirements for energy conservation and efficiency.

While most people can't get by with this type of vehicle alone, the fuel efficiency (up to 100 miles per gallon/42.4 kilometers per liter or more) can make scooter commuting a worthwhile investment, particularly if you live in a warmer climate.

- Carpool and/or combine trips whenever possible. The greatest amount of air pollution is released when an engine is cold. Combine errands on a single trip whenever possible. This helps vehicles run more efficiently and reduces the overall miles traveled.
- Maintain your vehicle properly. Regularly check the tire pressure and keep the engine tuned for optimal performance and fuel economy.
- Drive slower. Driving at speeds above 60 miles per hour (96.6 kilometers per hour) decreases fuel efficiency.
- Reduce equipment idling. If you will be sitting in your car for more than a few minutes, turn the engine off.
- Use mass transportation instead of individual vehicles. The fuel used per person is much lower for mass transportation than for travel in a single occupancy vehicle.
- Avoid unnecessary air travel. Airplanes are a significant source of carbon emissions, and their emissions are made worse by the fact that they occur high in the atmosphere. If possible, replace short-haul flights with travel by train or bus.

Energy Consumption by Appliances

According to the US Energy Information Administration, the average US household consumed 11,280 kilowatt hours of electricity in 2011. Where does all this energy go? The US Department of Energy has estimated how much electricity is required to run different appliances you may have in your home. Which appliance in your home requires the most energy to run? Considering the number of hours each appliance is used, which appliance uses the most energy overall? What is your biggest leverage point? Does it surprise you?

Appliance	Electricity Requirement
Fish aquariums	50 to 1,210 watts
Coffee maker	900 to 1,200 watts
Washer	350 to 500 watts
Dryer	1,800 to 5,000 watts
Dishwasher	1,200 to 2,400 watts
Ceiling fan	65 to 175 watts
Window fan	55 to 250 watts
House fan	240 to 750 watts
Hair dryer	1,200 to 1,875 watts
Clothes iron	1,000 to 1,800 watts
Microwave	750 to 1,100 watts
Average desktop PC	60 to 250 watts
Laptop	50 watts
Refrigerator	725 watts
Televisions	65 to 170 watts
Toaster	800 to 1,400 watts
DVD player	20 to 25 watts
Vacuum	1,000 to 1,440 watts
Water heater	4,500 to 5,500 watts

- Purchase locally produced goods and products. Many foods, particularly fruits and vegetables, are transported thousands of miles before they reach local stores. Shop at local farmers' markets or ask your grocer to stock locally produced foods to minimize the fuel used to transport goods.
- Move information, not people or things. Take advantage of the telephone and Internet as much as possible to avoid unnecessary travel. For example, rather than driving to several stores to find out which one carries a product or to compare prices, call the stores instead or research it on the Internet.

1.3.4 Rejecting, Reducing, Reusing, and Recycling Materials

The three Rs (reduce, reuse, and recycle) is a common phrase used to remember how to make greener choices. Reducing the materials used has a greater impact than reusing products or recycling them. This is true because using less of a material means less has to be produced, causing less impact from harvesting, manufacturing, and transportation. Likewise, reusing materials is better than using disposable materials that you recycle. Each reuse is one less product that has to be made. When you do use disposable goods, always try to recycle them instead of just throwing them out. Recycling is a good source of raw materials for new products. It is also an important way to save resources for the future.

The best choice of all, however, is a fourth R—reject. Be on the lookout for ways to reject the need to use raw materials in the first place. Rejecting the offer of a bag when you've only purchased one item is a good example. Most receipts can also be rejected—think twice before you say yes at the gas pump or ATM. From online statements for your bills to email instead of letters, there are many opportunities to reject the use of raw materials. Together, the four Rs—reject, reduce, reuse, and recycle—can help you remember the best way to be green in the choices you make every day.

1.3.5 Planting Vegetation

One important way to offset carbon emissions is by planting vegetation, such as trees or a garden. As mentioned earlier, plants absorb CO_2 and produce oxygen. They also provide shade, reduce stormwater runoff, promote groundwater recharge, and prevent soil erosion. There are many options for creating a beautiful and functional landscape while minimizing the impact of fertilizers,

pesticides, herbicides, and emissions from lawn equipment. These include the following:

- Plant native plants in your landscape instead of alien or invasive species. Native plants are well-adapted to local climate conditions and pests, and can often be grown without irrigation, pesticides, or fertilizer.
- Group plants with similar needs together in the landscape. This allows you to focus the application of irrigation, pesticides, and fertilizer only on the areas that really need it, while avoiding over-irrigation of other parts of the landscape.
- Plant edibles as part of your landscape, or even replace your lawn with a garden that can provide food for you and your family.
- Use manual equipment instead of gasoline-powered equipment for landscape maintenance. A standard walk-behind lawnmower puts out as much emissions as 11 cars, while the average riding mower generates as much pollution as 34 cars. Hybrid mowers are now available that use small amounts of gasoline to generate electricity that powers the mower. These mowers can also be used as generators to power electrical tools and equipment in the field.

1.3.6 Finding Better Energy Sources

After reducing your energy consumption as much as possible, investigate alternative sources of energy. These include distributed renewable energy systems such as wind turbines, hydropower, or photovoltaics you can install on site. Other options include purchasing green power if your utility provider offers it. This generally involves paying a small surcharge per kilowatt-hour to support the development of renewable energy resources such as large-scale wind, solar, and geothermal energy. Geothermal energy is derived from heat generated within the earth.

If green power is not available in your area, you can also reduce your carbon footprint through the purchase of carbon offsets. Carbon offsets can be purchased from certification companies and represent a certain amount of carbon that your purchase helps to reduce. For example, you may elect to purchase a carbon offset to mitigate the impacts of air travel, and you may even be asked if you'd like to buy an offset when you purchase a plane ticket. The money spent on offsets is used to help fund projects that reduce the amount of carbon emissions somewhere else. Carbon offsets can be used to plant trees, increase energy efficiency, or develop renewable energy sources. For example, a company may offset the carbon generated by its production processes by helping a school replace an old heating system. Spending money to reduce carbon emissions somewhere else is one way to be carbon neutral (that is, generating no greenhouse gases at all).

Think About It

Reducing Your Carbon Footprint

Compare your carbon footprint with the list of actions in this section. Which actions can you take to reduce your carbon footprint? Which ones seem easiest to achieve? Be sure to consider what kinds of changes you will be most likely to sustain over time. Remember that changes in technology are often easier to sustain than changes in behavior.

Did You Know?

Making Greener Choices

Even small choices have an impact. The carbon footprint of the average cheeseburger is 6.6 pounds (0.003 metric tons). This includes the energy required to grow the feed for the cattle, grow the vegetables, grow and process the grain, store and transport the components, and cook the burger.

70101-14_SA06.EPS

Additional Resources

Field Guide for Sustainable Construction., Department of Defense – Pentagon Renovation and Construction Office. (2004). PDF, 2.6 MB, 312 pgs. Available for download at **www.wbdg.org**.

Greening Federal Facilities, 2nd Ed. US Department of Energy Federal Energy Management Program. (2001). PDF, 2.1 MB, 211 pgs. Available for download at **www.wbdg.org**.

Natural Capitalism. Lovins, A., Hawkin, P., and Lovins, L.H. (1995). Little, Brown, & Company, Boston, MA. Available online at **www.natcap.org**.

Sustainable Buildings Technical Manual. Public Technologies, Inc./US Department of Energy. (2006). PDF, 3.1 MB, 292 pgs. Available for download at **www.greenbiz.com**.

Sustainable Buildings and Infrastructure: Paths to the Future. Pearce, A.R., Ahn, Y.H., and Hanmi Global. (2012). Routledge, London, UK.

Sustainable Construction: Green Building Design and Delivery, 3rd Ed. Kibert, C.J. (2012). Wiley, New York, NY.

The HOK Guidebook to Sustainable Design, 3rd Ed. Odell, W. and Lazarus, M.A. (2015). Wiley, New York, NY.

1.0.0 Section Review

1. Reducing your overall energy use is a way to
   ____ .
   a. improve health and comfort
   b. save trees
   c. improve the economy
   d. reduce your carbon footprint

2. The increase in greenhouse gases is a result of
   ____ .
   a. using biofuels
   b. burning fossil fuels
   c. using ethanol
   d. increased satellite use

3. The water used to flush toilets is commonly
   ____ .
   a. treated to the same quality standards as drinking water
   b. nonpotable
   c. treated with antibacterial chemicals
   d. treated with special salts

Section Two

2.0.0 Best Practices for Reducing Environmental Impacts of Projects

Objective

Identify technologies and practices that reduce environmental impacts of a project over its life cycle.

a. Describe the life cycle phases of a building and its impacts on the green environment.
b. Identify green site and landscape best practices and describe their pros and cons.
c. Identify green water and wastewater best practices and describe their pros and cons.
d. Identify green energy best practices and describe their pros and cons.
e. Identify green materials and waste best practices and describe their pros and cons.
f. Identify green indoor environment best practices and describe their pros and cons.
g. Identify green integrated strategies and describe the pros and cons of those alternatives.

Trade Terms

Absorptive finish: A surface finish that will absorb dust, particles, fumes, and sound.

Aerated autoclaved concrete (AAC): A lightweight, precast building material that provides structural strength, insulation, and fire resistance. Typical products include blocks, wall panels, and floor panels.

Aeration: Mixing air into a liquid substance.

Albedo: The extent to which an object reflects light from the sun. It is a ratio with values from 0 to 1. A value of 0 is dark (low albedo). A value of 1 is light (high albedo).

Alternative fuel: Any material or substance that can be used as a fuel other than conventional fossil fuels. They produce less pollution than fossil fuels. These include biodiesel and ethanol.

Best Management Practice (BMP): A way to accomplish something with the least amount of effort to achieve the best results. This is based on repeatable procedures that have proven themselves over time for large numbers of people.

Bio-based: A material derived from living matter, either plant or animal.

Biodegradable: Organic material such as plant and animal matter and other substances originating from living organisms. These are capable of being broken down into innocuous products by the action of microorganisms.

Biofuel: A solid, liquid, or gas fuel consisting of or derived from recently dead biological material. The most common source is plants.

Biomimicry: The act of imitating nature to create new solutions modeled after natural systems.

Bioswale: An engineered depression designed to accept and channel stormwater runoff. A bioswale uses natural methods to filter stormwater, such as vegetation and soil.

Blackout: A situation where the electrical grid fails and does not provide any power.

Blackwater: Water or sewage that contains fecal matter or sources of pathogens.

Brownfield: Property that contains the presence or potential presence of a hazardous substance, pollutant, or contaminant. Cleaning up and reinvesting in these properties reduces development pressures on undeveloped, open land and improves and protects the environment.

Brownout: A situation where the electrical grid provides less power than normal, but enough for some equipment to still work.

Building-integrated photovoltaics (BIPVs): Photovoltaic systems built into other types of building materials. See *photovoltaic*.

Byproducts: Another product derived from a manufacturing process or a chemical reaction. It is not the primary product or service being produced.

Charrette: An intense meeting of project participants that can quickly generate design solutions by integrating the abilities and interests of a diverse group of people.

Chlorofluorocarbons (CFCs): A set of chemical compounds that deplete ozone. They are widely used as solvents, coolants, and propellants in aerosols. These chemicals are the main cause of ozone depletion in the stratosphere.

Cob construction: An ancient building method using hand-formed lumps of earth mixed with sand and straw.

Commissioning: A review process conducted by a third party that involves a detailed design review, testing and balancing of systems, and system turnover.

Deconstruction: Taking a building apart with the intent of salvaging reusable materials.

Downcycling: Recycling one material into a material of lesser quality. An example is the recycling of plastics into lower-grade composites.

Endangered species: A population of a species at risk of becoming extinct. A threatened species is any species that is vulnerable to extinction in the near future.

Flood plain: An area surrounding a river or body of water that regularly floods within a given period of time.

Flushout: Using fresh air in the building HVAC system to remove contaminants from the building.

FSC certified: Wood or wood products that have met the Forest Stewardship Council's tracking process for sustainable harvest.

Fugitive emissions: Pollutants released to the air other than those from stacks or vents. They are often due to equipment leaks, evaporative processes, and wind disturbances.

Graywater: A non-industrial wastewater generated from domestic processes including laundry and bathing. Graywater comprises 50 to 80 percent of residential wastewater.

Green Seal Certified:: A certification of a product to indicate its environmental friendliness. Green Seal is a group that works with manufacturers, industry sectors, purchasing groups, and government at all levels to green the production and purchasing chain. Founded in 1989, Green Seal provides science-based environmental certification standards.

Grid intertie: A connection between a local source of power and the utility power grid.

Hardscape: Paved areas surrounding a project, including parking lots and sidewalks.

Hydrologic cycle: The circulation and conservation of Earth's water supply. The process has five phases: condensation, infiltration, runoff, evaporation, and precipitation.

Indoor air quality (IAQ): The content of interior air that could affect health and comfort of building occupants.

Insulating concrete form (ICF):: Rigid forms that hold concrete in place during curing and remain in place afterwards. The forms serve as thermal insulation for concrete walls.

Integrative design: A collaborative design methodology that emphasizes the input of knowledge from several areas in the development of a complete design.

Infiltration: The movement of air from outside a building to inside through cracks or openings in the building envelope.

Just-in-time delivery: A material delivery strategy that reduces material inventory. The material that is delivered is used immediately.

Lithium ion (Li-ion): A type of rechargeable battery in which a lithium ion moves between the anode and cathode. They are commonly used in consumer electronics.

Local materials: Materials that come from within a certain number of miles from the project. Materials produced locally use less energy during transportation to the site. The LEED system of building certification offers points for the use of regional or local materials.

Maintainability: An indication of how difficult it is to properly maintain a building system or technology.

Massing: The three-dimensional shape of a building, including length, width, and height. Determines the amount of a building exposed to sunlight.

Minimum efficiency reporting value (MERV): A measurement scale designed in 1987 by the American Society of Heating, Refrigeration, and Air-Conditioning Engineers (ASHRAE) to rate the effectiveness of air filters. The scale is designed to represent the worst-case performance of a filter when dealing with particles in the range of 0.3 to 10 microns.

Monofill: A landfill that accepts only one type of waste, typically in bales.

Multi-function material: A material that can be used to perform more than one function in a facility.

Nano-material: A material with features smaller than a micron in at least one dimension.

Nickel cadmium (NiCad): A popular type of rechargeable battery using nickel oxide hydroxide and metallic cadmium as electrodes.

Nonrenewable: A material or energy source that cannot be replenished within a reasonable period.

Nontoxic: Substances that are not poisonous.

Offgassing: The evaporation of volatile chemicals at normal atmospheric pressure. Building materials release chemicals into the air through evaporation.

Papercrete: A fiber-cement material that uses waste paper for fiber.

Passive solar design: Designing a building to use the sun's energy for lighting, heating, and cooling with minimal additional inputs of energy.

Passive survivability: The ability of a building to continue to offer basic function and habitability after a loss of infrastructure (for example, water and power).

Pathogen: An infectious agent that causes disease or illness.

Peak shaving: An energy management strategy that reduces demand during peak times of the day and shifts it to off-peak times, such as at night.

Pervious concrete: A mixture of coarse aggregate, Portland cement, water, and little to no sand. It has a 15 to 25 percent void structure and allows 3 to 8 gallons of water per minute to pass through each square foot. Also known as permeable concrete.

Phase change material (PCM): A material that stores or releases large amounts of energy when it changes between solid, liquid, or gas forms.

Photosensor: A sensor that measures daylight and adjusts artificial lighting to save energy.

Pollution prevention: The prevention or reduction of pollution at the source.

Post-consumer: Products made out of material that has been used by the end consumer and then collected for recycling.

Post-industrial/pre-consumer: Material diverted from the waste stream during the manufacturing process.

Radio-frequency identification (RFID): An automatic identification method. It relies on storing and remotely retrieving data using devices called RFID tags.

Rainwater harvesting: The gathering and storing of rainwater.

Rammed earth: An ancient building technique similar to adobe using soil that is mostly clay and sand. The difference is that the material is compressed or tamped into place, usually with forms that create very flat vertical surfaces.

Rapidly renewable: A material that is replenished by natural processes at a rate comparable to its rate of consumption. The LEED system of building certification rewards use of rapidly renewable materials that regenerate in 10 years or less, such as bamboo, cork, wool, and straw.

Recyclable: Material that still has useful physical or chemical properties after serving its original purpose. It can be reused or remanufactured into additional products. Plastic, paper, glass, used oil, and aluminum cans are examples of recyclable materials.

Recycled content: A material containing components that would otherwise have been discarded.

Recycled plastic lumber (RPL): A wood-like product made from recovered plastic, either by itself or mixed with other materials. It can be used as a substitute for concrete, wood, and metals.

Reusable: A material that can be used again without reprocessing. This can be for its original purpose or for a new purpose.

Salvaged: A used material that is saved from destruction or waste by being reused.

Sick Building Syndrome: A variety of illnesses thought to be caused by poor indoor air. Symptoms include headaches, fatigue, and other problems that increase with continued exposure.

Smart material: Materials that have one or more properties that can be significantly changed. These changes are driven by external stimuli.

Softscape: The area surrounding a building that contains plantings, lawn, and other vegetated areas.

Solid waste:: Products and materials discarded after use in homes, businesses, restaurants, schools, industrial plants, or elsewhere.

Solvent-based: A material that consists of particles suspended or dissolved in a solvent. A solvent is any substance that will dissolve another. Solvent-based building materials typically use chemicals other than water as their solvent, including toluene and turpentine, with hazardous health effects.

Strawbale construction: A building method that uses straw as the structural element, insulation, or both. It has advantages over some conventional building systems because of its cost and availability.

Structural insulated panel (SIP): A composite building material used for exterior building envelopes. It consists of a sandwich of two layers of structural board with an insulating layer of foam in between. The board is usually oriented strand board (OSB) and the foam can be polystyrene, soy-based foam, urethane, or even compressed straw.

Sustainably harvested: A method of harvesting a material from a natural ecosystem without damaging the ability of the ecosystem to continue to produce the material indefinitely.

Takeback: A condition where manufacturers recover waste from packaging or products after use, at the end of their life cycle.

Thermal mass: A property of a material related to density that allows it to absorb heat from a heat source, and then release it slowly. Common materials used to provide thermal mass include adobe, mud, stones, or even tanks of water.

Urea formaldehyde: A transparent thermosetting resin or plastic. It is made from urea and formaldehyde heated in the presence of a mild base. Urea formaldehyde has negative effects on human health when allowed to offgas or burn.

Vapor-resistant: A material that resists the flow of water vapor.

Virgin material: A material that has not been previously used or consumed. It also has not been subjected to processing. See *raw material*.

Volatile organic compounds (VOCs): Gases that are emitted over time from certain solids or liquids. Concentrations of many VOCs are up to 10 times higher indoors than outdoors. Examples include paints and lacquers, paint strippers, cleaning supplies, pesticides, building materials, and furnishings.

Walk-off mat: Mats in entry areas that capture dirt and other particles.

Water-based: Materials that uses water as a solvent or vehicle of application.

Waterproof: A material that is impervious to or unaffected by water.

Water-resistant: A material that hinders the penetration of water.

Wetland: Lands where saturation with water is the dominant factor. This determines the way soil develops and the types of plant and animal communities living in the soil and on its surface.

Xeriscaping: Landscaping that requires little or no water for irrigation. Xeriscaping can be achieved through smart plant selection, mulching, and other tactics.

Zoning: Rules for placing similar items next to one another, as in plants within a landscape, or types of buildings within a community.

Throughout recorded history, humans have constructed buildings to protect themselves and their possessions. Buildings provide shelter from adverse climate conditions such as rain, snow, wind, and temperature extremes. They also offer privacy and security. In addition to these roles, built facilities also serve the following purposes:

- Collection, treatment, and/or storage of solid, liquid, and gaseous waste
- Provision and distribution of pure water
- Processing and distribution of agricultural products into food
- Manufacturing and distribution of various products

The impacts of buildings on the environment have not always been obvious, but the effect over time is undeniable. Buildings are responsible for over 10 percent of the world's freshwater withdrawals. They also use 25 percent of the world's wood harvest and nearly 40 percent of its material and energy flows.

In the United States, 72 percent of all electricity use is related to building construction and operation. Almost a third of all new and remodeled buildings suffer from poor indoor air quality (IAQ) due to emissions from various sources, volatile organic compounds (VOCs) released from building products, and pathogens from inadequate moisture protection and ventilation. Poor indoor air results in variety of illnesses known collectively as Sick Building Syndrome, which is thought to cost more than $60 billion annually in lost productivity and medical costs nationwide.

Nearly one-quarter of all ozone-depleting chlorofluorocarbons (CFCs) are emitted by building air conditioners. The processes used to manufacture building materials also contribute. Approximately half of the CFCs produced around the world are used in buildings. This includes refrigeration and air conditioning systems and fire extinguishing systems. CFCs are also in certain insulation materials. In addition, half of the world's fossil fuel consumption is attributed to the servicing of buildings. Lighting accounts for 20 to 25 percent of the electricity used in the United States annually. Offices in the US spend 30 to 40 cents of every energy dollar for lighting. This makes it one of the most expensive and wasteful building features. Finally, the construction industry is responsible for 20 to 40 percent of the total municipal solid waste stream.

2.1.0 Facility Life Cycle

The life cycle of a built facility may range from 30 years to over 100 years. It typically runs through the following phases:

- *Planning or pre-design* – A facility's life starts with an idea during the planning or pre-design phase. This phase focuses on defining the function of the building. The physical requirements of the building are determined. Goals for the budget and construction schedule are set. Legal or regulatory constraints are identified that should be taken into account during design. The outcome of the planning phase is typically a set of requirements for the facility. These describe the functional expectations the owner has for the facility.

- *Design* – The second phase of the facility life cycle is design. This is when the facility is transformed from an idea into a set of construction drawings and specifications. A design meeting called a **charrette** may be used to obtain input from project stakeholders.
- *Construction* – During construction, workers follow the set of construction documents to construct a building that meets the owner's requirements. The outcome of the construction phase is a completely functional building.
- *Operation and maintenance* – After construction, the operation and maintenance phase of the life cycle begins. The building is used to meet the owner's needs. This is typically the longest phase of the life cycle. Operation is the process by which the facility performs its intended functions. Maintenance consists of all actions performed on the facility to keep it in good operating condition. It includes activities such as changing light bulbs, servicing the building systems, and cleaning the facility. It also includes minor repairs or replacement of building components that break down.
- *Rehabilitation or end of life cycle and disposal* – At some point, a facility will no longer meet the owner's requirements. A possible choice is to rehabilitate or reconstruct the facility to improve its performance. This can be even more challenging than the original construction. Another possibility is to end the life cycle of the facility through **deconstruction** or demolition. Deconstruction is a planned, careful disassembly of the facility to salvage building components for future use. Demolition is a more destructive process.

Built facilities affect the green environment over their life cycles. *Figure 13* shows some of the relationships between a building and the environment during its life cycle. Note the import and export of materials, energy, and waste, all of which have impacts on the green environment.

Think About It

Inventory the World Around You

Consider the built environment in which you're sitting right now. What features of the building do you think work well in terms of green performance? Which features do you think could be better? Working individually or in small teams, explore the building in which you are taking this class. Also, consider the site on which it is located. Begin by drawing a floor plan for the building. Add significant features of the site. Take an inventory of the following items and mark the location of each observation on your floor plan or site map. For each item, list why it's an example of excellence or how it could be improved.

- *Energy use* – Look around your building for technologies and features that save or waste energy. Pay special attention to the building envelope, which is the outside skin of the building including windows and doors. Examine the heating, ventilation, and cooling systems. Study the lighting in the building. What opportunities do you see to improve the energy performance of the building? What features already work well?
- *Water use* – Inspect your building for features that save or waste water. Pay special attention to faucets, fixtures, appliances, landscaping, and cooling systems that rely on water. What opportunities can you find to improve the water performance of your building? What things already work well?
- *Materials* – Now consider your building from the standpoint of materials. Examine the structure of the building, its enclosure, and interior and exterior finishes. Look at how the building is currently being used. What opportunities do you see to improve the performance of your building in terms of the materials used for construction or consumed for operations? What things are already satisfactory?
- *Indoor environment* – Look around the interior spaces of your classroom building. What about the spaces contributes to a comfortable and productive environment? What could be better? Pay special attention to lighting, acoustics, views, and indoor air quality.
- *Outdoor environment* – Step outside your building and inspect the site on which it is located. How could the building and its site be improved from the outside? In which ways is it already satisfactory? Pay special attention to the landscape (both the plants and the paved areas). Also consider the location of the building itself with regard to the people who use it. What transportation options exist now? What could be better?

When you've finished with your inventory, share your findings with the rest of the class. Have other students noticed things you didn't see?

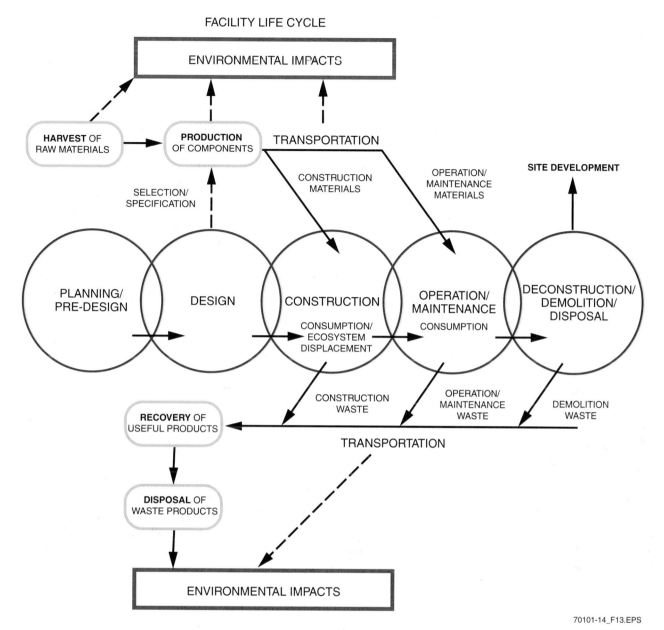

FACILITY LIFE CYCLE

Figure 13 Interaction between a building and the green environment.

You can change your actions to improve those impacts. Some impacts are due to the materials and energy used by the facility while other impacts come from the waste streams that leave the facility. How the facility interacts with the site on which it is located creates impacts, as well as how the facility interacts with its users. The following **best management practices (BMPs)** represent a spectrum of options for improving the impacts of the built environment on the green environment:

• *Site and landscape* – This category involves choosing a good site for a facility. It involves the specific placement of the facility on the site and the design of landscaping. Avoiding damage to the site is important. Restoring the quality of the site after construction is also important.

• *Water and wastewater* – This category addresses optimizing the use of water. It also covers alternative water sources and wastewater sinks.

• *Energy* – This category addresses optimizing the use of energy. It also covers methods of using energy more efficiently. It includes balancing energy demands and seeking alternative sources for energy.

• *Materials* – This category includes optimizing the use of materials. It promotes using abundant, renewable, or multifunction materials. It also involves seeking alternative sources for materials.

- *Waste* – This category includes eliminating or preventing waste. It involves reusing waste within the facility system, sharing it with other systems, and storing it for future use.
- *Indoor environment* – This category includes preventing problems at the source, segregating polluters, taking advantage of natural forces, and giving users control over their environment.
- *Integrated strategies* – This category includes practices that result in multiple benefits from a single action. It includes capitalizing on construction means and methods. It also discusses making technologies do more than one thing and exploiting relationships between systems.

Each of these categories is shown in *Figure 14* and described in detail in the following sections.

2.2.0 Site and Landscape Best Practices

The first category of best practices deals with the facility site and landscape. The most important decision for a building is the selection of a site, followed by where to put the building on the site. These two decisions affect the building's performance over its life cycle. Next, the development of the site should use low-impact principles and features as well as ensure that ecosystems on the site are restored to their best quality.

2.2.1 Site Selection

The choice of site affects the building's energy use and environmental impact, and it also governs travel to and from the facility. For example, choosing a site in an already developed area provides building occupants with access to existing stores and services. It may even allow them to take advantage of walking, bicycling, or public transit to access the building instead of having to drive a car. A good location can also provide a positive contribution to an existing neighborhood. Choosing a site that is a brownfield is a possibility. A brownfield is a site that may have real or perceived environmental contamination. Cleaning up a brownfield represents a positive step for the community, and it may provide a tax credit or development incentive for the builder.

Avoid sites with valuable ecological resources or higher levels of risk from environmental damage, such as sites with wetlands or habitats of threatened or endangered species. The goal is to preserve these resources and prevent the need to mitigate or help correct any impacts. It also saves considerable expense. Choose sites that are well out of flood plains or areas where mudslides or wildfires are common. Be sure to consider the long-term risks of resource depletion as well. For example, the water supply in some areas is becoming scarce. This may mean restrictions on how facilities are developed and used in the future.

Some sites can share common resources with other sites if the facilities have different hours of operation. For example, a bank that is only open during the day may allow parking for a nightclub or restaurant at night. This eliminates the need for two separate parking lots. Choosing sites in already developed areas provides access to the existing infrastructure, which includes streets as well as water, power, and sewer lines. This can save time and money and reduce the negative effects on the site.

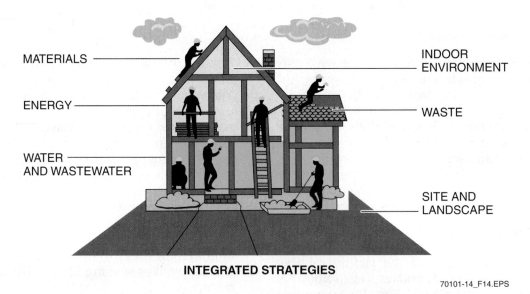

Figure 14 Categories of best practices.

2.2.2 Building Orientation

A building's relationship to the sun has a large impact on energy use. In fact, it can save 30 to 40 percent in heating and cooling costs over the life cycle of the building. *Figure 15* shows a building in relation to the sun's path in the Northern Hemisphere. The building is oriented with its long axis running from east to west, and most of the windows are on the south side of the building. In the summer, the sun passes higher in the sky and the overhangs on the building help shade the windows, which keeps the building cooler. In the winter, the sun passes lower in the sky. This allows sunlight to enter the windows under the overhangs and provides free heat and light. This is an example of passive solar design. Passive solar design takes advantage of the sun's energy to provide heat without using electrical or combustion energy. Another important idea in passive solar design is building massing, where the shape and size of the building are designed to provide access for solar energy to enter the building to provide heat, power, and daylight. Massing and orientation can also influence a building's ventilation. If oriented with openings to take advantage of prevailing winds, buildings can be naturally ventilated in some climates.

Trees can also be used to provide free cooling. Deciduous trees on the south, east, and west sides of a building can provide shade in the summer, which reduces cooling costs. In the winter, they lose their leaves and allow solar energy to reach the building. This energy provides warmth and heat gain in the winter. Trees also provide shelter from prevailing winds, which helps to reduce the heating load on the building. Preserve existing trees whenever possible. In addition to energy benefits, trees also have positive effects on water retention, air quality, and property values.

Another important consideration when placing the building on site is avoiding areas that are difficult to develop. For example, areas with steep slopes are more difficult to develop than flat areas, and require more earthwork and/or larger foundations. The same is true for areas with high water tables. After examination of the entire site, thought must be given to both initial and long-term costs of the building location.

2.2.3 Site Development

Site development involves changes made to the landscape of the site. These changes may adjust terrain, change vegetation, and add amenities for humans. Ideally, site development should avoid disturbing existing landscape, especially on sites that have not been previously developed. Contractors may set up fences to protect trees and other important landscape during construction. Damage to site ecosystems can be avoided with careful planning of the construction process. Limiting heavy equipment to areas where it is needed will also help to protect soil and avoid damage to vegetation.

Heavy equipment used for site development contributes to a project's carbon footprint. Heavy equipment should be used as efficiently as possible

THE SUN'S PATH IN THE SKY

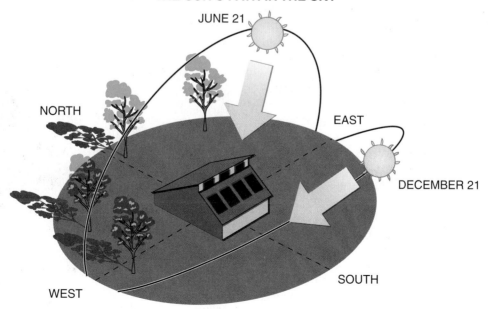

JUNE 21

NORTH

EAST

DECEMBER 21

WEST

SOUTH

70101-14_F15.EPS

Figure 15 Building oriented to take advantage of the sun's path.

for earthwork. More efficient equipment can also be used. Heavy equipment that uses biodiesel instead of fossil fuels, including excavators and trucks can be found at some sites. Hybrid vehicles are available that combine combustion engines with electric motors and batteries. These vehicles can also reduce the carbon impacts of equipment idling.

Landscaping is divided into two main categories: softscape and hardscape. Softscape refers to the vegetated parts of a site, such as grass, trees, and bushes. Hardscape refers to areas that are paved or otherwise developed. Best practices associated with low-impact landscaping include the following:

- *Native plants* – Use native plant species. Native plants are well adapted to the climate and require little or no irrigation, fertilizer, or pesticides once established. Avoid exotic plants from other parts of the world. They often out-compete native plants and can become invasive. Kudzu is a well-known invasive species common in the southeast United States. Many common plants, such as English ivy, are also invasive and should be avoided. Visit **www.plants.usda.gov** and click on the links for noxious and invasive plants. This website also has information on native plants that may do well in your area.

- *Zoned landscaping* – Group plants with similar needs, which is called zoning. Zoned landscaping allows you to focus irrigation, fertilizer, and pesticides only on the plants that require it. It can also help reduce the risk of wildfire around buildings. This is accomplished by keeping areas near buildings clear of flammable vegetation and debris.

- *Mulching* – Use bark chips, pine straw, or other natural materials to suppress weeds and retain moisture. This reduces the need for irrigation and pesticides, and helps stabilize new plants until they become established.

- *Xeriscaping* – Xeriscaping is the selection of native plants that require no irrigation once they are established. It is common in the southwest, where xeriscape areas of rock, sand, and native plants replace green lawns that would require high volumes of water in order to survive.

- *Runoff control* – Use vegetated areas to absorb and treat stormwater runoff from paved areas. Bioswales and rain gardens are engineered basins that collect stormwater. They are similar in purpose except that rain gardens tend to have a more formal layout of plants and stones, while bioswales are planted using a more natural approach. Once collected, the water is percolated slowly back into the soil (*Figure 16*). Plants within the bioswale also help to remove contaminants

from runoff. These pollutants would otherwise have to be treated at a wastewater treatment plant to avoid contaminating local water streams.

- *Permeable pavement* – Permeable pavement captures stormwater and allows it to percolate back into the soil (*Figure 17*). Options include **pervious concrete** (concrete that contains spaces through which water can move), stabilized soil, and grid systems. Stormwater that drains over pavement may become contaminated and require collection and treatment. This is particularly true in areas where vehicles drive or park. These contaminants include engine leaks, rubber and asbestos particles, and other residues. When concentrated in runoff, these contaminants can pollute local streams and rivers. Stormwater runoff can also cause stream damage after running across hot

70101-14_F16.EPS

Figure 16 A bioswale receives stormwater runoff from a parking lot.

70101-14_F17.EPS

Figure 17 Permeable paving options include pervious concrete and grid systems that allow water to pass into the soil.

pavement, leading to thermal pollution of local streams, which can harm aquatic plants and animals. Some permeable pavements provide basic treatment of contaminants, accomplished through plant roots or microorganisms in the pavement pores.

- *Light-colored pavement* – Pavement that is light in color helps minimize the amount of heat absorbed from the sun. Light colors have high albedo values. The term *albedo* describes the extent to which an object reflects light from the sun. It is a ratio with values from 0 to 1; a value of 0 is dark (low albedo), while a value of 1 is light (high albedo). High-albedo pavements help to minimize the urban heat island effect, which is caused in part by dark asphalt and buildings that absorb, rather than reflect, heat. Using light-colored concrete can reduce overall temperatures during summer months and lower the air conditioning requirements of surrounding buildings. *Figure 18* shows high- and low-albedo pavements.
- *Alternative transportation* – Some landscaping features promote the use of alternative transportation (*Figure 19*). These include sheltered transit stops, bike racks, and bike paths and

HIGH-ALBEDO   LOW-ALBEDO

70101-14_F18.EPS

Figure 18 High- and low-albedo pavements.

sidewalks or pedestrian trails. Setting aside special parking for carpools or alternative fuel vehicles rewards drivers for making greener choices. One type of alternative fuel is biofuel, which is commonly made from plants.

2.2.4 Restoring Ecosystems

After the project is built, restore natural ecosystems to the extent possible by setting aside areas of the site to remain undeveloped. Native species require little or no maintenance; they provide landscape beauty and serve as a natural habitat for various creatures, including birds and butterflies. Open spaces also provide a place for building occupants to be outside and enjoy nature. Whenever possible, work with adjoining sites to create connected undeveloped areas, including local streams and waterways.

2.3.0 Water and Wastewater Best Practices

The second category of best practices involves the sources and uses of water for a facility. These practices also deal with sinks for a building's wastewater. Nearly all buildings require a source of water to meet the needs of occupants. This includes water for drinking, washing, and waste disposal.

Water may appear to be abundant, but less than 3 percent of all water on Earth is fresh water. The rest is salt water, which is mostly unusable. Of the small amount of fresh water available, 69 percent is trapped as ice and snow cover and 30 percent is stored as groundwater. Less than 1 percent is available on the surface as freshwater lakes and rivers. The hydrologic cycle continuously replenishes the freshwater supply. This process has five phases: condensation, infiltration, runoff, evaporation, and precipitation. In many areas, the rate of use exceeds the recharge rate, leading to aquifer depletion.

To conserve water, the most important action you can take is to eliminate unnecessary uses of it. Then increase the efficiency of the water you do use. After determining how to reduce the need for water as much as possible, the next steps are to look for alternative sources of water and seek alternative ways to remove and treat wastewater.

2.3.1 Reducing Water Use

If you are involved with an existing building, complete a water audit. This will help find leaks and identify opportunities for improvement.

70101-14_F19.EPS

Figure 19 Amenities to promote alternative transportation.

Your local water authority may be able to provide information on completing a water audit.

Another way to eliminate water use is to install waterless toilets and urinals. Waterless urinals (*Figure 20*) take advantage of the fact that urine is a liquid, meaning it does not require water to be conveyed through wastewater pipes. Waterless urinals use a special trap that allows urine to pass through, but prevents sewer gases from escaping.

Waterless toilets are also available. For example, composting and incinerating toilets both work without the use of water. Composting toilets

70101-14_F20.EPS

Figure 20 Waterless urinal.

convert human waste into a useful product that can be used to improve the soil. Incinerating toilets use electrical energy or other fuel to convert the waste into ash. Incinerating toilets use a considerable amount of energy, which should be taken into account as a tradeoff against the water savings.

2.3.2 Optimizing Water Use

Many types of fixtures are available to reduce water use in showers and sinks. These include low-flow or aerated fixtures and alternate types of controls. Aeration is the introduction of air into the water flow. It is a common technique to reduce the actual flow of water in faucets and showerheads while maintaining the perception of high-volume flow. It works well for hand washing and showers, but can be frustrating when filling a pot with water for cooking or cleaning. Some aerated kitchen faucets allow the user to control the degree of aeration and restore full flow when needed.

Another option is foot-operated sinks. These allow precise control of flow timing while leaving hands free. More precise automated flow controls are also becoming available for sinks and toilets. The EPA's Water-Sense® program provides a labeling system to help identify low-flow fixtures and controls (see **www.epa.gov/watersense**).

Toilet technology has improved considerably over the last decade. There are now many options available to improve the efficiency of water use. In addition to conventional low-flow toilets, several manufacturers now offer dual-flush toilets. These allow a full-volume flush for solid waste and a half-volume flush for liquids. Automated dual-flush control units are also available. These units dispense either a full-volume or low-volume flush depending on how long the user is within range of the flush sensor. Manual controls are

Waterless Urinals: Balancing Water Savings with Maintenance Requirements

Waterless urinals save significant water over the life of a building. They eliminate water used to flush liquid waste down sewer pipes. However, there are tradeoffs to think about in choosing a urinal that will work well. Waterless urinals require a different maintenance routine than usual. Proper training is required for maintenance staff to keep waterless urinals working properly.

There are two main types of waterless urinals. One type relies on a special cartridge to prevent sewer gases from escaping, which can be costly and the cartridge must be periodically replaced. The second type uses an oil or gel to provide a seal that must be replenished regularly. The gel is less expensive than the cartridge. However, calcium deposits may build up in the waste pipes with this technology, and a plumber must then periodically remove these deposits. This adds to the maintenance cost of the urinal.

Choosing the best urinal for a project requires thinking about operations and maintenance. What resources will be available? What training will be provided? Choosing a urinal that does not fit with maintenance routines or expectations may result in a need to replace it in the future. It may even require changes in walls to provide water supply pipes for replacement urinals. Think carefully about the whole life cycle of a product before using it, to avoid problems later.

also included to allow users to control the flush level if desired.

When selecting water sources for landscaping, consider drip irrigation or irrigation controlled by moisture sensors and timers. Drip irrigation provides a slow release of water directly to the root zone of the plant. This results in less evaporation than a typical sprinkler system. Moisture sensors and timers control the operating periods of irrigation systems. This ensures that water is only dispensed when needed. It also allows watering during the early morning when water has time to reach plant roots before evaporating. With either system, be sure to properly locate and maintain all components of the system to avoid waste.

Water-efficient appliances, such as high-efficiency dishwashers and washing machines, reduce water use by up to 50 percent over traditional units. Since these units use less hot water, they also save energy and are reviewed under the EPA's ENERGY STAR program. See **www.energystar. gov** for a list of washing machines and dishwashers that meet ENERGY STAR criteria.

2.3.3 Finding Alternative Sources of Water

A third option for greening your use of water is to seek alternative sources of water. Focus on uses that do not require water to be treated to drinking water standards. Much of the water used in homes and businesses does not require this high level of water quality. Think about the extra energy and treatment required to use drinking water to flush a toilet.

One alternative source of fresh water is known as **rainwater harvesting**, which involves capturing and using rainwater. Rainwater harvesting has been used on a small scale since development began. Larger scale systems have been used in arid parts of the country for decades and are likely to become more common as water scarcity increases. *Figure 21* shows a rainwater harvesting system. These systems require a collection surface, typically a roof that is relatively debris-free. The surface must also be made from a chemically stable substance. Older metal roofs should not be used as they may have joints sealed with lead-based solder, which could contaminate the water. Asphalt shingles are also not recommended since they tend to shed particles as they age. Slate, synthetic, and lead-free metal roofs are all good collection surfaces.

In addition to the collection surface, rainwater systems require components to provide filtration, storage, and overflow. Most existing roof drainage systems can be retrofitted for rainwater harvesting. Depending on rainfall levels, a roof can capture tens of thousands of gallons of water per year.

Another alternative source of water is **graywater**. This is used water from sinks, showers, and laundry facilities. It is called graywater because it contains a small amount of contamination that can be safely treated on site. Full-building graywater systems collect and filter water for irrigation or toilet flushing (*Figure 22*). Water recovered for reuse is transmitted through purple-colored pipes instead of regular supply pipes. Graywater systems must be carefully designed to meet code requirements, as graywater supports bacterial growth while in storage. Provisions must be made to balance

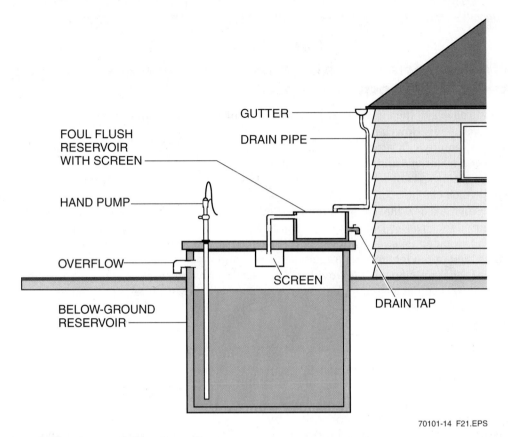

FOUL FLUSH RESERVOIR WITH SCREEN

HAND PUMP

OVERFLOW

BELOW-GROUND RESERVOIR

GUTTER

DRAIN PIPE

SCREEN

DRAIN TAP

70101-14 F21.EPS

Figure 21 Components of a rainwater harvesting system.

supply with demand and allow overflow into a sewer when necessary. Graywater that remains in the system after a specified period must be disposed of rather than reused.

Small-scale graywater systems are also available. They capture water from one fixture, such as a shower, and divert it for use in another fixture, such as a toilet. Some systems fit under a bathroom sink and work well for retrofits.

Water from toilets and dishwashers is considered blackwater. This is because it contains pathogens. The pathogens are from either human waste or from animal fats associated with dishwashing. Blackwater requires separate piping and additional levels of treatment before reuse.

2.3.4 Finding Alternative Sinks for Wastewater

The final opportunity to green a building's water systems involves finding other uses or alternative treatments for wastewater. A building's wastewater stream also includes stormwater. Stormwater may be contaminated by materials from parking lots and roofs, and is also likely to be much hotter than local water streams. Alternative systems are available to treat this type of wastewater.

Graywater systems can capture water that has not been heavily contaminated. This water can be

redirected for certain uses without treatment if allowed by local codes. These systems can also be the first step in a more comprehensive system. Lightly contaminated water is recycled for heavier use, and then treated on site using a separate system.

One type of alternative wastewater technology is a plant-based constructed wetland. Constructed wetlands have been used in a variety of climates and applications. *Figure 23* shows a constructed wetland for a small town with about 900 residents. This series of ponds is filled with plants that provide increasing levels of water treatment. A primary treatment step removes solids from the water stream before it enters the ponds. After treatment, the clean water is discharged to the local river.

Smaller plant-based or bio-based wastewater treatment systems can be used to treat the wastewater from a single building. One example of a bio-based system is called a Living Machine®. Living Machines® combine plants, bacteria, and other organisms into a simulated ecosystem that uses wastewater as a nutrient source. *Figure 24* shows a bio-based wastewater treatment system at the Rocky Mountain Institute in Snowmass, Colorado. This system includes hedgehogs and lizards as part of the indoor ecosystem, and it even produces bananas in winter.

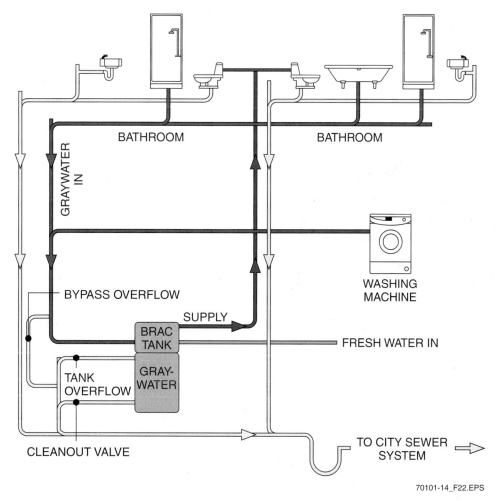

70101-14_F22.EPS

Figure 22 Components of a building-scale graywater system.

2.4.0 Energy Best Practices

The energy used in homes and buildings has considerable impacts on the green environment. As described in Section One, producing and using energy has a strong effect on the carbon footprint of our society. Many construction activities rely on carbon-intensive energy sources such as fuel for heavy equipment. The resulting buildings also consume energy over their extended life cycle. Much of that energy is still produced from non-renewable energy sources, creating greenhouse gases. Renewable energy sources represent only a small share of all energy used. Energy best practices can be used to reduce the impacts of buildings on climate change and resource depletion, including the following:

- Avoiding unnecessary energy use
- Optimizing energy use
- Balancing electrical loads
- Seeking alternative energy sources

Generators provide power where it is needed when electrical connections are not available. To do so, most generators rely on combustion of fossil fuels. This combustion results in air pollutants as well as noise pollution on the job site. Extension cords are also required to connect equipment, which can create trip hazards and unsafe conditions. However, generators allow continuous production without the need to stop and change batteries.

Think About It
Powering Construction Activities

Advances in battery technology have made construction easier in many ways. No longer are fossil-fuel generators required for power on sites where electricity has not yet been installed. But what is the impact of batteries on the natural and human environments compared to generators? Which is greener?

70101-14_F23.EPS

Figure 23 Constructed wetland wastewater treatment system.

70101-14_F24.EPS

Figure 24 Bio-based wastewater treatment system.

Batteries allow workers to move quickly and easily around the site to get their work done. However, they must be recharged and they eventually wear out. Multiple batteries are needed to allow workers to keep working while other batteries are charging, and charging batteries at the job site may still require generators. The time it takes to change batteries can reduce productivity, although risks from long power cords are eliminated. Manufacturing batteries has high environmental impacts—they use toxic heavy metals and are energy-intensive. Their disposal can also be a problem, but they should always be properly recycled.

While some battery technologies are greener than others, not all batteries work equally well for all uses. New generators that use solar power are also becoming available. Choosing the best approach for a task requires considering tradeoffs and planning. This ensures that the best choice for health, safety, and the environment is made while not compromising productivity.

2.4.1 Avoiding Unneeded Energy Use

The best way to green a building's energy use is to reduce or eliminate the demand for energy. One approach involves setting back the thermostat and wearing a sweater in the winter or opening windows to take advantage of the breeze in the summer. Another is to include timers and occupancy sensors that can be used to control the lighting or temperature when a space is unoccupied.

Other strategies reduce demand with no changes required in occupant behavior. These include passive heating and cooling, daylighting, and heat recovery ventilators (HRVs). HRVs capture waste heat from exhaust air and use it to preheat incoming air. Some systems also equalize humidity in a similar way. Other systems take advantage of the thermal mass of building materials. Thermal mass is a property that allows a material to absorb heat and then release it slowly. For example, adobe houses have thick walls that absorb heat during the day and release it at night.

Another simple strategy is a high-albedo roof (*Figure 25*). These roof materials and coatings reduce heat gain by reflecting solar energy rather than absorbing it. As *Figure 25(B)* shows, these surfaces need not be white and shiny. New coatings are available that reduce heat gain even with darker color roofs. In many cases, the albedo of a roof can be increased without cost by choosing a different roof color during design.

(A) HIGH-ALBEDO MEMBRANE ROOF

(B) HIGH-ALBEDO METAL ROOF

70101-14_F25.EPS

Figure 25 High-albedo roofs.

- Increasing insulation in wall cavities and attics
- Sealing cracks and using vapor-resistant barriers to reduce infiltration
- Installing high-efficiency windows and doors
- Using low-emissivity paint or radiant barriers in attics to reduce heat gain

WARNING!
Make sure to always use proper fall protection when working at height. High albedo roofs may help a building be greener, but they have also been reported to have higher glare. This can endanger workers installing and working on them. Wear proper protection from the sun, including eye protection and sunscreen, while working on high albedo roofs. Be aware of roof edges that may be hard to see. Be sure to tie off properly.

High-performance building envelopes can also eliminate unnecessary uses of energy for heating and cooling. A building's envelope is the outer shell that protects it from the elements. If you added up the area of all cracks and openings in the building envelope of a typical American house, it would equal about one square yard (nearly one square meter). That is the equivalent of leaving a window fully open year-round, even when the heater or air conditioner is on. Ways to improve a building's envelope include the following:

All of these methods can reduce the demands on facility space conditioning equipment and increase occupant comfort. Commissioning the building envelope is also important for high-performance envelopes. Commissioning has traditionally been limited to building mechanical and electrical systems. However, having a third party check that all parts of the envelope are installed correctly and working together can be valuable to prevent future problems.

Finally, use good control systems for lighting to minimize the use of energy during hours when daylight provides enough light. Dual switching

Choosing a Roof:
Balancing Energy Savings with Durability

High albedo roofs can help save significant energy over the life of a building by keeping the roof cooler. However, there are tradeoffs to think about with some types of high albedo roofs. According to the Lawrence Berkeley Laboratory, the dirt that accumulates on a high albedo roof can reduce energy savings by 20 percent or more. This means that the roof must be cleaned regularly to maintain its function. Roof cleaning requires working at height, which poses additional safety risks.

Keeping roof systems cooler can help a roof last longer due to reduced stress from expansion and contraction. However, certain types of high albedo roof membranes have a much shorter service life than conventional roofs. For instance, thermoplastic polyolefin (TPO) roofs are more environmentally friendly in many ways than other roof membranes. However, their service life may be significantly less than a conventional membrane roof. Many factors must be considered in choosing a roof system. The design and details of the roof will play a large role in roof service life. Other factors include embodied energy, use of toxic materials, recyclability, and cost. Commitment to maintenance is also important.

involves wiring light switches so that combinations of lamps can be switched on or off depending on how much light is needed. This helps users to adjust the light to appropriate levels. Properly labeling the switches allows for easier use.

2.4.2 Optimizing Energy Use

There are many opportunities for optimizing energy use. One way is proper sizing of heating, ventilation, and air conditioning (HVAC) equipment. Oversized equipment operates at peak efficiency only a few days per year. A better choice is the use of two smaller units. Under normal conditions, only one unit is required and it operates at peak capacity. Both units are used together to meet peak demands that occur only a few hours per year. The energy savings offsets the initial cost of two systems. Having two units also means one can be kept running while working on the other, increasing maintainability.

Proper operation and maintenance is an often-overlooked way to save energy. Even simple maintenance, such as cleaning the lenses on lighting fixtures, can greatly increase light output and reduce the need for supplemental lighting. Continuous commissioning can be applied to existing buildings. Commissioning is a procedure to examine each building component to ensure it is operating as originally designed. It can be used to identify ways to save energy with existing equipment. Continuous commissioning typically pays back in less than two years. Operator training is also important to ensure that building equipment is used properly.

Energy audits can identify ways to retrofit existing systems with more efficient ones. A lighting retrofit has rapid payback in many buildings (*Figure 26*). Replace older magnetic ballasts with electronic ballasts in fluorescent lights. You can convert existing T-12 fixtures to T-8 when changing ballasts. Other lighting retrofits include installing solid state LED lighting fixtures or T-5 fluorescent lighting. These retrofits require replacement, not just of lamps and ballasts, but also of fixtures themselves. Maintenance requirements can be reduced due to the much longer service life of solid state lighting. Adding photosensors or occupancy sensors is another lighting retrofit, as they can shut off lights when they are not needed.

70101-14_F26.EPS

Figure 26 Lighting retrofits are an easy way to save energy.

Prevention Through Design

One of the most important parts of a green building is its envelope. This system provides a boundary between outside and inside. It allows control of the climate inside the building to make occupants comfortable. It also can affect the building's energy and water performance in important ways. From high albedo roofs that reflect unwanted heat, to roof- or wall-mounted photovoltaics or solar heating systems, to vegetated roofs and walls, the building envelope requires careful design and construction.

70101-14_SA08.EPS

It is particularly important to consider construction and maintenance of these innovative systems during design. Many injuries that occur on green projects are associated with falls from height while working on the building envelope. Prevention through Design (PtD) is a practice of thinking ahead during design about the ways in which people could be hurt while working on the building. The building is then designed to eliminate danger when possible, or provide ways to reduce that danger. For instance, all of these roofs require regular maintenance during operation. An access door could be added to the design to avoid the need for ladders, and tieoff points or walking surfaces may be included on the roof to make safe access easier.

PtD is especially important in green buildings, given the additional functions played by the building envelope. Not only should designers consider safety issues in the design, they should also consider maintainability in choosing building systems. Maintainability of a system refers to the difficulty associated with proper maintenance of that system. Owners must be prepared for different maintenance requirements that go along with innovative technology. Without proper maintenance, many innovative systems will not work correctly. This can cause problems during operation. Designers should discuss maintenance requirements with owners when choosing building technologies. Owners must ensure that proper training and resources are required to keep the building working properly over time.

WARNING!
Fluorescent lamps contain mercury, and they must be properly disposed as hazardous waste. Old ballasts and thermostats also often contain toxic components and must be treated as hazardous waste.

WARNING!
Lighting retrofits require working at height. Be sure to tie off and use appropriate safety measures.

Replacing tank-type water heaters with tankless models is another way to save energy. Tankless models cost more initially, but they eliminate standing heat loss since water is only heated as needed. This saves energy over the life cycle. Regular maintenance is important to keep both tank-type and tankless water heaters working properly.

High-performance HVAC systems also save energy. Variable speed fans, pumps, and motors enable these systems to operate at optimum levels. This makes the system cycle on and off less frequently, which increases user comfort and saves energy. Optimized distribution systems also help reduce heating, cooling, and ventilation costs. Sealing and insulating ducts is an important way to ensure that the energy used to condition air actually benefits the users of the building.

Finally, ultra high-efficiency appliances and equipment can quickly pay for themselves in energy savings. Look for ENERGY STAR appliances when making replacements or buying new equipment.

2.4.3 Balancing Electrical Loads

The electrical grid is a complex system of power plants and distribution networks. The end systems that consume the power are also complex. To keep the grid functioning, energy supply must remain balanced with demand. If this does not happen, the system becomes unstable. When demand exceeds supply, additional generators are brought online to meet the additional need. If these peak generators cannot meet demand, the system may experience brownouts, which are noticeable losses in system voltage that may damage equipment such as motors. In the worst case, the system becomes so unstable that safety mechanisms activate to take parts of the grid offline. This is known as a blackout.

Load balancing is a way to limit the loads placed on the power grid. This allows power plants for peak power generation to be smaller, and reduces the need for new power plants. The highest demand for energy is typically during the middle of the day in the hottest part of summer. At this time, industry is operating at maximum output and commercial buildings are operating air conditioning at maximum capacity. Shutting down unnecessary equipment during these periods is one way to reduce the need to bring peak generators online.

Peak shaving reduces demand during peak times of the day and shifts demand to off-peak times, such as at night. Utilities often charge lower rates during off-peak times. This provides an incentive for users to shift demands to these periods. Thermal storage systems are designed to take advantage of peak shaving. Energy is used at night to super-cool a thermal storage medium. This medium then absorbs heat to provide cooling during the day.

Energy management systems use electronic controls to balance loads within buildings during peak periods. They do this by reducing the amount of power sent to equipment that is tolerant of voltage variations, including certain types of air conditioners, pumps, fans, and motors. They also maintain a steady stream of power to equipment that requires a constant voltage input, such as computers. Depending on the local or regional climate for power production, the price for electricity during peak hours can be quite high. In many areas, an investment in energy management equipment can rapidly pay for itself in terms of reduced energy costs. Often, utility companies will finance or invest in these systems for built facilities. They may also provide technical assistance to support implementation.

Finally, information can be a useful way to balance loads and shave peak demand. Large institutions such as universities often pay for power based on time-of-day usage. During the hottest parts of the year, rates can easily triple when demand for power is highest. Some institutions send out email reminders to shut down unnecessary equipment and lighting during these periods (*Figure 27*). This saves both electrical energy and money by reducing use during high-price periods.

2.4.4 Finding Alternative Energy Sources

After reducing demand, optimizing efficiency, and balancing loads, the last step is to explore alternative sources of energy. There are two primary ways to obtain alternative energy: buying energy

---

GOING GREEN

Reducing the Cost of the Brooklyn Bridge

The Brooklyn Bridge has gone green with LED lights, part of New York City's plan to save energy and cut greenhouse gases. The 160 LED fixtures each use only 24 watts of power instead of the 100-watt mercury bulbs formerly in place, and last three times as long as the mercury vapor lamps they replaced. This saves energy costs and reduces carbon dioxide emissions by about 24 tons each year.

70101-14_SA09.EPS

WE NEED YOUR HELP!

Energy Advisory

Electricity Price Caution is in effect today for the time period from 3 pm to 7 pm.
Electricity Price Caution is issued when the electricity prices are 3 to 6 times higher than normal.

Here are some actions you can take to help reduce our electricity consumption:
1. Activate the energy saving or "sleep" mode on computers and copiers.
2. Turn off your computer monitor when you are away from your desk for more than 15 minutes.
3. Turn off lights when out of your office or cubicle.
4. Turn off lights in unused common areas such as copy rooms, break rooms, conference rooms, unoccupied rooms, and restrooms.
5. If you have control over the thermostat setting for the air conditioner, raise it by two degrees during the peak hours. Consider raising the level of the thermostat further when your facilities are unoccupied.
6. Shut off nonessential machinery, computers, and other equipment.
7. Consider reducing the number of copiers available for use during peak hours.

XYZ Power Company – Prices for Today

Hours	cents/kWh
1:00	2.1203
2:00	2.0594
3:00	2.0235
4:00	2.0263
5:00	2.0278
6:00	2.0213
7:00	2.1258
8:00	2.1257
9:00	2.1637
10:00	2.2745
11:00	2.4378
12:00	3.3367
13:00	4.9845
14:00	8.0461
15:00	11.3748
16:00	12.5938
17:00	12.4085
18:00	11.2554
19:00	8.9338
20:00	7.2654
21:00	6.3741
22:00	4.9265
23:00	3.3319
24:00	2.2695

Average price = 6.0211 per hour at end interval

70101-14_F27.EPS

Figure 27 Using information to encourage energy conservation for peak shaving.

from a green power provider, or generating power on site. Green power providers generate electricity from renewable energy sources such as wind power and solar power. They avoid nonrenewable power sources such as fossil fuels. To see if certified green power is available in your area, visit the Green-e website at **www.green-e.org.** Green-e is a nonprofit organization that certifies the renewable energy power companies sell to consumers. *Figure* 28 shows the Green-e Energy Certified logo that companies can use on the certified renewable energy products they sell.

The second option is to explore the possibility of on-site renewable energy. This includes photovoltaics, wind turbines, gas-fired microturbines, and fuel cells. *Figure 29* shows a roof-mounted wind turbine providing power to a commercial building.

Smart Grid

Smart grid technology allows buildings to communicate with power producers to better manage power needs. Smart grid technology also gathers information on system performance and alerts utilities to problems that require maintenance or repair. A key part of the smart grid is automation technology that allows utilities to control power-consuming equipment in a building from a remote location.

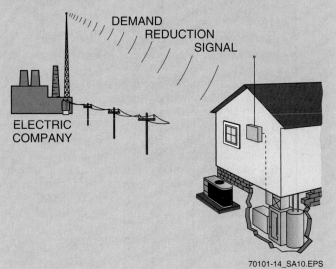

70101-14_SA10.EPS

70101-14_F28.EPS

Figure 28 The Green-e Energy Certified power provider logo.

Standalone power systems require costly and high-maintenance battery storage for excess power. Electrical energy is difficult to store. The most effective way to use on-site power generation is with a grid intertie, which is a connection to the local utility grid. It allows you to route excess power back to the grid when you generate more than you can use. You can also continue to buy power from the utility when your on-site system does not meet your demands. With photovoltaic systems, peak capacity falls at the same time of day as peak demand, which is in the afternoon. In some areas, you may even be able to sell your power. If you live in a part of the country where there is not much excess generating capacity, utilities may help finance your project. They may also provide technical assistance to set up renewable energy systems. Check online at **www.dsireusa.org** to see what programs are available in your area.

WARNING!
Photovoltaic panels generate electrical current whenever they are exposed to light, whether or not they are connected to the electrical system of a building. Be aware of this risk and take proper precautions – they are still potentially dangerous even if they have been disconnected.

WARNING!
Battery banks used for power backup and storage pose both electrical and chemical hazards. Many batteries use lead-acid technology, both of which are toxic chemicals. Batteries may also produce hydrogen gas that needs to be properly vented outdoors.

With demand for energy only likely to grow in the future, you need to be aware of opportunities to manage consumption of electrical power in built facilities. Energy investments save money over the life cycle of a facility and can increase the ability of the facility to withstand fluctuations in power supply. This reduces vulnerability to natural disasters and other threats. As dependence

GOING GREEN

Rebuilding the World Trade Center

Developers of the new World Trade Center complex are committed to ensuring that it meets green goals. Each of the seven towers will be developed to meet or exceed LEED Silver standards. Sustainable design has been used in other parts of the site as well, including the WTC Transportation Hub, Memorial, and vehicle security center. Green features include water-efficient landscaping, local and recycled materials, and high-performance systems for indoor air quality and comfort. The 7 WTC Tower was the first green commercial office building in New York City, certified LEED Gold in 2006.

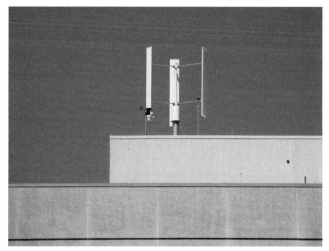

70101-14_F29.EPS

Figure 29 Small-scale wind turbine on a commercial building.

on power grows, the vulnerability of power systems will increase as well. Facilities with well-managed, minimal energy requirements will be the least vulnerable to fluctuations in power (and prices) from utility suppliers. These facilities will use on-site power generation as well as passive systems for lighting, heating, and cooling.

2.5.0 Materials and Waste Best Practices

The materials used to construct and maintain buildings are one of the biggest contributors to their impact on the green environment. As mentioned earlier, buildings are responsible for 25 percent of the world's wood harvest. Current rates of deforestation amount to an area the size of the state of Georgia each year. Buildings also account for about 40 percent of global material and energy flows. That adds up to over three billion tons each year. Each product or material used in a building has a life history that includes the following:

- Harvesting of all the raw materials required to make it
- All the byproducts generated during its production
- All the energy used to transport its components from place to place

In addition to raw materials, solid waste also contributes to a building's impact on the green environment. Buildings are responsible for up to 20 to 40 percent of the solid waste stream in some parts of the United States.

Greening your resource use starts by eliminating the unnecessary use of new materials. This includes using materials that come from waste, salvaged, or recycled sources. It also includes re-using and adapting existing buildings instead of building new ones.

The second step is to use materials more efficiently. Construction methods that reduce waste contribute to this goal. A whole new generation of multi-function materials is now available that provide structure, enclosure, insulation, and other functions. Using these products reduces the amount of raw material needed to meet building needs and save on shipping and packaging. These materials are pre-engineered to minimize waste on site.

Seeking better sources for the materials we use can help ensure an ongoing supply of products to meet future needs. Switching to abundant, renewable materials helps to preserve the limited supply of nonrenewable resources. Ensuring that materials are sustainably harvested means that the supply of those materials can continue

GOING GREEN

Wind Farms

The US Department of Energy states that wind farms may be a major source of power as we approach the year 2030. Their analysis shows that up to 20 percent of US power needs could be handled by wind power, which would reduce pollution to the same extent as taking 140 million cars off the road.

70101-14_SA12.EPS

indefinitely. Using local materials helps to reduce the impacts of transportation of raw materials and supports local economies.

Finally, finding better sinks (destinations) for waste from the building helps the green environment. Some of the ways to improve performance in this area include on-site and off-site recycling and using biodegradable or reusable packaging. Salvaging or deconstructing building systems at the end of their lives is also green. Biodegradable packaging is made of natural materials that readily break down using natural processes. Some manufacturers sponsor takeback programs that recover a product or its packaging at the end of its life cycle.

2.5.1 Eliminating the Unnecessary Use of New Materials

The first step in greening building materials is to eliminate the unnecessary use of new materials. This means finding ways to get the job done with fewer materials. It also means substituting materials or parts of materials that aren't new whenever possible. Reusing materials is a green choice. Every product that is salvaged and reused saves

energy and raw materials. Recycled content prevents the need to harvest virgin materials, which helps conserve valuable resources.

Perhaps the greenest choice of all is to reuse or adapt an existing building instead of building a new one. Not only do you eliminate the need for many new materials, you also reduce site development and infrastructure costs. Be aware that existing buildings present a challenge, as certain conditions are not always as expected. Sometimes contamination from lead-based paint, asbestos, or other substances must be remediated. Some systems may need to be replaced or upgraded as well. Avoid reusing windows, older HVAC equipment and appliances, and older plumbing fixtures; they are likely to be very inefficient and may no longer meet current standards.

Pay attention to packaging when procuring raw materials. In some cases, you can specify that products be delivered with reusable packaging that can be returned to the manufacturer. In other cases, packaging is recyclable or even biodegradable. If product packaging is bio-based (that is, made from substances derived from living matter), look for opportunities to compost or chip it on site for use as a soil amendment. This also

Think About It

Multi-Function Materials: Structural Insulated Panels

Multi-function materials, such as structural insulated panels (SIP), are becoming more common in the industry. This is due to both time savings in the construction schedule, and energy savings during building operation. These materials replace several different building materials with a single product. What materials does a SIP replace? How does a SIP change the way a building is built? Which trades are affected, and how? What tasks are easier? What tasks are more difficult? Why?

70101-14_SA13.EPS

reduces the need to purchase new materials for this purpose.

Salvaged materials can be a very green contribution to a project. If an existing building is being demolished, look for ways to reuse its materials in the new building. Often, concrete and masonry rubble can be reused for fill, subbase for pavements, or drainage. Timber in good condition can often be reused for structural purposes, or remilled for flooring or siding. Masonry units may be reused as well. Materials that cannot be reused on the new project may be worth salvaging for other projects. These may include doors, hardware, and various fixtures.

New materials with recycled content also reduce the use of virgin materials. Common materials with recycled content include steel and concrete. Recycled materials include a wide variety of finishes, structural materials, and even landscaping materials (*Figure 30*). With materials like steel and concrete, you may not even be able to tell the difference from virgin materials. Other products may include recycled content as composites. Recycled plastic lumber (RPL) is one such product. RPL combines post-consumer or post-industrial plastic waste with wood fibers

or other materials. The result is an extremely durable wood substitute.

Recycling reduces the amount of waste sent to landfills and creates a source of raw materials for new products. However, not all recycling is the same. True recycling means using the waste to create the same kind of product. Many materials can only be downcycled. They are turned into something else that can never be turned back into the original product. One example is plastics recycling. Recycling plastic into clear new containers is difficult due to mixing colors and contamination during recycling. Chemical bonds can also decay in the plastic. Most recycled plastics are turned into composite products such as recycled plastic lumber, lower-grade plastic bags, or even carpet backing. Recycling plastics does reduce the amount of new materials required, but at a loss of both purity and quality. The best way to be green is to reduce the use of these products in the first place.

Recycled content can come from a variety of sources. Some recycled content is produced as a byproduct of manufacturing processes. This material is called post-industrial/pre-consumer recycled content. It can be recovered and reused without ever leaving the factory. Other recycled

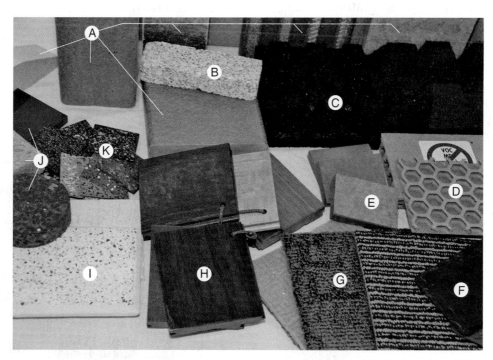

A. Recycled plastic lumber
B. Insulating concrete form material with recycled polystyrene pellets
C. Recycled rubber turf stabilizer
D. Recycled paper fiber board
E. Recycled paper fiber countertop
F. Recycled rubber shingles

G. Recycled content carpet (face fiber and backing)
H. Salvaged lumber
I. Recycled paper and glass cement tiles
J. Recycled glass countertops
K. Recycled rubber flooring

70101-14_F30.EPS

Figure 30 Waste-based and recycled content materials.

content comes from products that have been produced and used by consumers, then recovered for recycling. This material is called *post-consumer* recycled content. Post-industrial recycled content results from waste or inefficiency in manufacturing; rather than recycling this material, it would be better to improve the manufacturing process. Post-consumer content is better than post-industrial recycled content because it truly closes the material loop. Products made from post-consumer waste are valued more highly when certifying a green building than pre-consumer waste.

Pollution prevention is the careful design of products and processes to eliminate the use or waste of materials. One example is to eliminate finishes and expose structural materials instead. For example, concrete floors can be stained, polished, or textured to create beautiful surfaces that do not require additional finishes, such as wood or tile. Exposed ceilings (*Figure 31*) eliminate the need for extensive dropped ceiling systems and provide visual interest while saving costs.

In many projects, natural systems can be used to perform the functions provided by engineered systems at lower cost. One example is living machines made of plants, snails, bacteria, and other organisms. Living machines, natural drainage swales, and constructed wetlands can be used to collect and treat wastewater without requiring any input of chemicals. The result is healthy plants and purified water that can be safely used for other purposes. Some of these practices cost more than conventional systems. However, you can often save enough money to pay for the

70101-14_F31.EPS

Figure 31 Exposed ceilings eliminate unneeded materials.

investment through savings in other systems and/or life cycle cost savings.

Many building strategies also take advantage of the lessons offered by the green environment. This approach is known as **biomimicry**—imitating nature to create new solutions modeled after natural systems. One example is African termite mounds that maintain constant internal temperatures even with large fluctuations in outdoor air temperature. Buildings have been designed to mimic these mounds using natural ventilation. The mounds use an intricate network of tunnels to control the flow of air for optimum quality, temperature, and moisture levels. These self-sufficient mounds also include a type of farming.

Think About It

Recycling vs. Downcycling

This photo shows two different products with recycled content: a carpet tile and a piece of industrial flooring. Which one do you think is recycled? Which one is downcycled?

70101-14_SA14.EPS

The termites supply fungus in the mounds with chewed wood fiber, which the fungus then breaks down into a usable food source. Nature produces solutions with optimum efficiency because plants and animals that cannot compete simply do not survive. As the world heads into an age of increasingly scarce resources and more people who need them, biomimicry offers valuable lessons for buildings and technologies alike.

2.5.2 Using Materials More Efficiently

The second strategy for greening materials is to use materials more efficiently, which means getting more benefit from the materials you do use and/or using fewer materials to achieve the same result. One way to do this is to use multi-function materials. Multi-function materials do more than one thing as part of a building. They can speed up construction, save building materials, and reduce waste. Every system and technology used in a project comes with overhead, which includes extra packaging, transportation, and other costs that do not add value to the product itself. When one product serves as multiple systems, the overhead for that product is a fraction of the overhead associated with the multiple systems it replaces.

Many multi-function materials can serve as the structure of the building. These include **insulating concrete forms (ICFs)**, **structural insulated panels (SIPs)**, and **aerated autoclaved concrete (AAC)**. Each of these structural systems (*Figure 32*) also serves as the building enclosure, insulation system, and mounting surface for interior and exterior finishes. By using these products, construction of one system achieves as much as three or four traditional building systems. This saves time, labor, packaging, transportation, and money.

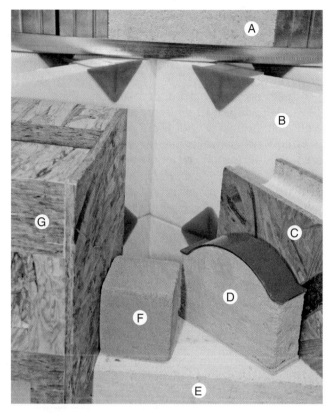

A. Embedded steel stud SIP
B. Polystyrene insulating concrete form
C. Polystyrene stress skin SIP
D. Aerated glass stress skin SIP
E. Aerated autoclaved concrete
F. Fiber-reinforced aerated concrete
G. Straw core stress skin SIP

70101-14_F32.EPS

Figure 32 Multi-function materials.

An energy-related example is **building-integrated photovoltaics (BIPVs)**. These solar panels are built into components that play other roles in a building. Some BIPVs are used to generate

Biomimicry

GOING GREEN

Scientists have begun to look to nature for inspiration as they develop new building materials and systems. One example is a spider web. Spiders produce fibers stronger than Kevlar and spin them into self-healing structures that can resist tremendous loads. Spider webs contain many lessons for structural materials and design.

70101-14_SA15.EPS

electricity when the sun shines on a window, and others are manufactured as solar shingles—they generate power while acting as the roof itself. Still others are used as shading devices for windows or parking lots.

Green roofs are another example (*Figure 33*). These vegetated roofs have energy benefits as well as benefits for roof life.

Green roofs are typically installed on top of a roof membrane. They can be of varying complexity, and help to ballast the roof itself. They also protect the membrane from solar radiation, which would otherwise cause the material to break down. Green roofs stabilize temperatures on the roof, reducing stress on the roof membrane and keeping the building cooler. They help to absorb rainfall, which reduces stormwater runoff. They are also a microhabitat for birds, insects, and plants, and they also help to clean the air. This single system, although it may cost more than a traditional roof, provides significant benefits that can make it a good investment. All benefits and costs must be considered when deciding what types of systems to use.

> **WARNING!**
> Installing and maintaining a green roof requires working at height, often by trades that are unfamiliar with fall protection. Green roofs should include design features to facilitate safe access for maintenance. Be sure to employ active fall protection when working on green roofs.

Optimal value engineered (OVE) framing for light wood frame construction is another way to use materials more efficiently. OVE framing is a green practice because it reduces the amount of wood needed. It also allows more insulation to be incorporated in the walls of the building (*Figure 34*). This reduces the amount of heat that can escape through the walls and increases the energy performance of the building.

OVE framing also saves considerable materials during construction by using wood more efficiently. This means less lumber, fewer cuts, and less waste. Key principles of OVE framing include:

- Designing buildings on 2-foot (0.6-meter) modules to make the best use of common sheet sizes
- Spacing wall studs, roof joists, and rafters up to 24" (0.6 meters) on center
- Using in-line framing where floor, wall, and roof framing members are vertically in line with one another and loads are transferred directly downward

(A)

(B)

70101-14_F33.EPS

Figure 33 (A) Green roofs provide enclosure, stormwater management, and thermal control while cleaning the air and providing a habitat for birds and insects. (B) Walls can also be vegetated.

- Using two-stud corner framing and inexpensive drywall clips or scrap lumber for drywall backing
- Eliminating headers in walls that are not loadbearing
- Using single lumber headers and top plates when appropriate

---

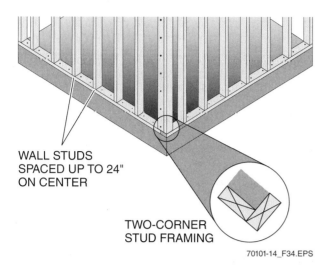

WALL STUDS
SPACED UP TO 24"
ON CENTER

TWO-CORNER
STUD FRAMING

70101-14_F34.EPS

Figure 34 Framing details for OVE wood frame construction.

These techniques reduce the amount of wood used and create a better thermal envelope. They also make construction easier for crews that must install plumbing, electrical, and HVAC services in the walls. Fewer structural members mean less drilling during installation and less effort to install. All of these attributes make OVE framing a good strategy to green your building and save both first costs and life cycle costs.

Another approach is to use smart materials. Smart materials work by changing in response to environmental conditions. For example, smart windows may automatically become more opaque in response to higher levels of sunlight. Smart window blinds may automatically adjust to follow the sun's path. Some materials are under development that can change colors or other properties as well.

Phase change materials (PCMs) are one class of smart building materials. These materials are capable of storing and releasing large amounts of energy when they change between solid, liquid, or gas forms. When the material reaches a temperature at which it changes phase, it can absorb or release a large amount of heat at an almost constant temperature. Materials at their phase change temperatures can be used to absorb or release heat to cool or heat a building with far less energy than conventional cooling or heating systems.

A special class of smart materials is called nano-materials. Nano-materials use extremely small particles that add functionality to building materials. For example, nano-coatings can be used to make paint repel dirt or concrete absorb air pollution. Nano-coatings are also being used to add mold resistance or biocidal (germ killing) properties to finishes. A new generation of photovoltaic panels are being developed using carbon nano-tubes, which improve efficiency by increasing the surface area exposed to the sun.

New generations of lightweight modularized construction systems are now available that can also do more with less. Prefabricated, factory-assembled building components reduce waste. They use materials more efficiently since they are manufactured in a controlled environment. Regular production allows manufacturers to optimize the use of materials and produce the most efficient components possible.

Think About It

OVE Framing: Impact on Cost

How does an improvement to one system change the requirements for another system? OVE framing is a good practice for saving lumber and improving insulation of walls. However, using more widely spaced studs changes how other materials like drywall are used. With fewer attachment points, regular drywall may flex, causing uneven wall surfaces that are difficult to finish. Thicker drywall is generally required to do a good job with wider stud spacing. This means that while the cost of framing may decrease, the cost of drywall will increase. What other impacts might occur? How would this change the process of assembling a wall?

70101-14_SA16.EPS

Modular construction components include carpet tile (*Figure 35*), raised floor systems, and demountable furniture systems. These systems allow rapid reconfiguration to meet changing user needs. They also permit replacement on a unit-by-unit basis in the case of damage. This means avoiding the need to replace entire rooms of carpet for one stained or worn area. Even regular materials can be used for modular construction, by using regular dimensions that match the standard sizes of building materials. For example, don't design decking that is 9 feet (2.7 meters) long, as decking comes in increments of 2 feet (0.6 meters). Expand your deck to match standard material sizes and avoid having to waste material.

Choose systems and materials of appropriate durability. If you know that the facility you're building is going to be remodeled every 5 to 7 years, don't use granite countertops unless you plan to recover and reuse them somewhere else. This idea is known as *design for disassembly*. Put your systems together in a way that you can take them apart at the end of their service life for possible reuse.

During the construction process, use lean construction methods to maximize efficiency. Lean construction eliminates unnecessary steps and materials, and eliminates waste by changing the way things are built. For example, just-in-time delivery of materials to the site means that products are installed right off the truck, not needing to be stored or staged. This eliminates the opportunity for materials to become damaged in storage. It also minimizes the opportunity for loss or damage as products are moved around the site.

Careful material tracking helps to avoid waste. A new method of material tracking involves the use of radio-frequency identification (RFID) tags. These tags make it easy to locate materials in crowded staging areas. If materials must be stored before installation, protect them from moisture and possible damage. Centralized cutting operations also help to increase material efficiency. Keep cutoffs in the same area where all cutting is done. This makes workers more likely to use them instead of pulling a new piece of material from the supply stack.

2.5.3 Finding Better Sources of Materials

Another way to green your use of materials is to find better sources. For example, rapidly renewable materials have become common. Rapidly renewable materials are a special class of bio-based materials that grow quickly. According to the US Green Building Council (USGBC)®, rapidly renewable materials are any material that can be sustainably harvested on less than a 10-year cycle (*Figure 36*). They may be derived from either plant or animal sources, including bamboo, cellulose fiber, and wool. Other examples are cotton insulation, corn-based carpet, blown soy insulation, agrifiber, linoleum, wheatboard, strawboard, and cork.

Other materials can be considered green because they are abundant. This includes buildings made from soil, such as rammed earth, adobe, or cob construction. Cob construction is an ancient building method using hand-formed lumps of earth mixed with sand and straw. Like cob, adobe typically includes plant fibers to help stabilize the soil. Other earth-based construction methods strengthen the soil by using small quantities of cement or lime. An innovative material in this category is papercrete, a fiber-cement that uses waste paper for fiber.

70101-14_F35.EPS

Figure 35 Carpet tile is a modular material that prevents waste.

What's Wrong with This Picture?

70101-14_SA17.EPS

A. Soy-based foam insulation
B. Wool carpet
C. Corn-based carpet fiber
D. Coir/straw soil stabilization mats
E. Corn-based carpet backing
F. Coconut wood flooring
G. Wheat straw fiberboard

H. Sorghum board
I. Bamboo flooring/plywood
J. Linoleum
K. Hemp fiber fabric
L. Cork flooring
M. Cotton fiber batt insulation

70101-14_F36.EPS

Figure 36 Bio-based and rapidly renewable materials.

Strawbale construction uses both abundant and bio-based materials. The bales of straw used for this purpose are typically three to five times more densely packed than agricultural bales. Strawbale construction can be either structural or used as fill with a wood frame or other structure. Bales are stabilized using rebar or bamboo spikes. Bond beams are used to ensure a load-bearing surface along the tops of walls. Strawbale is becoming more widely used in certain parts of the United States; Arizona even has a special section of the building code for this construction method.

Proper detailing to prevent moisture problems is critical for success. Deep overhangs and moisture barriers between bales and foundations are common. Strawbale construction has a high insulating value, and has a two-hour or greater fire rating depending on the exterior and interior finish.

Recycled content materials are also a better source than virgin materials for many applications. Post-consumer recycled content is especially desirable. However, many recycled content materials are composites. This means that they are combinations of multiple different types of raw materials. The prospects for recycling composite materials are limited. However, some composite materials outperform conventional materials, requiring lower maintenance and fewer raw materials.

People often confuse the term *recyclable* with *recycled*, but there's a big difference between the two. *Recyclable* means that the product itself can be recycled after it is used; *Recycled* means that a product contains material that has been previously used, either in industry during manufacturing (post-industrial) or by consumers (post-consumer). Post-industrial recycled content is less desirable than post-consumer recycled content, as post-industrial recycled content often results from wasteful manufacturing processes that could be improved. Scientific Certification Systems is an organization that provides certification of the recycled content of materials. Look for the SCS logo on products to be sure that their recycled content has been verified (*Figure 37*).

Look for sustainably harvested materials. The Forest Stewardship Council (FSC) is an organization that evaluates wood products to determine if they are sustainably harvested. Look for **FSC certified** wood products (*Figure 38*). More information on the FSC certification standards is available online at **www.fsc.org**.

Figure 37 SCS Certified content label.

Figure 38 FSC Certified wood product.

Seek out local sources of materials wherever possible for your projects. Using locally produced materials reduces the energy required to transport them. This is one of the biggest impacts of building material. Often, locally harvested materials can be obtained at lesser cost than imported materials. Using local materials can save time for procurement and helps to support the local economy.

2.5.4 Finding Better Sinks for Waste Streams

The final step you can take to green your materials use is to find better uses for building waste. Depending on the type of project you are building, multiple options may be available. Consult salvage companies for projects that involve the removal or rehabilitation of existing buildings. This will ensure that all reusable materials are removed before demolition begins. Older buildings may be good candidates for deconstruction, which is the

systematic disassembly of a building in the reverse order from how it was built. This process salvages timber and other materials for reuse.

Both salvaged materials from existing buildings and leftover materials from new projects can be donated to local nonprofit organizations involved with construction. Habitat for Humanity is a nonprofit organization that operates ReStores in various cities (*Figure 39*). ReStores are retail outlets that sell donated building materials. Companies that donate materials to a ReStore can take a tax deduction for their charitable contribution. At the same time, the ReStore obtains a source of materials that it can sell to raise money to build homes for needy families. If your project is likely to have usable materials left over, contact your local ReStore to ask about on-site pickups and donation opportunities. You can also shop at the ReStore yourself if you need small quantities of a certain item or would like to find salvaged or used materials for your next project.

For waste that is not suitable for donation, consider recycling as an option for disposal. Metal recycling is available throughout the United States. Organizations that recycle cardboard, drywall, ceiling tiles, carpet, concrete, masonry, and asphalt shingles are becoming more common. Consider contracting with a waste recovery company to simplify recycling. Depending on the waste streams generated, these companies deliver dumpsters at different times during the project. They may also provide worker training or housekeeping services to ensure that materials are sorted correctly. Some areas also have facilities for off-site sorting of co-mingled construction and demolition waste. These facilities take your waste at a reduced rate and sort it to recover

Figure 39 Habitat for Humanity ReStores accept donations of construction materials.

recyclable materials. Depending on market conditions and the composition of your waste stream, they may even pay you for your waste. In addition to typical building materials, batteries can also be recycled. **Lithium ion (Li-ion)** and **nickel cadmium (NiCad)** batteries are both rechargeable and recyclable.

> **WARNING!**
> On-site sorting of construction waste requires additional handling of material to be sure it is placed in the proper container. Sites with multiple waste containers can be more congested and pose hazards for workers. Additional handling of scrap and waste can also pose physical hazards such as cuts or strains. Plan site layout and logistics carefully. Use proper lifting practices and wear appropriate personal protective equipment when handling waste.

Bio-based waste is a good candidate for onsite chipping, mulching, or composting. Pallets, wood waste, cardboard, and even drywall can be chipped or mulched on site by specialty contractors. The result is a product that can be used as a soil amendment or for landscaping purposes. Some contractors also grind concrete and masonry rubble for use on site as a fill or subbase.

> **NOTE**
> Cardboard is a higher value waste that should be recycled rather than composted if possible.

> **WARNING!**
> Do not burn pressure-treated wood. Chemicals in the wood can become airborne and inhaled into the lungs. Always separate treated wood from wood being chipped for mulch. The chemicals used in pressure-treated lumber can contaminate the soil. Collect scraps and place in a dumpster reserved for pressure-treated material. Wash your hands thoroughly after handling pressure-treated material.

Certain kinds of waste, if produced in large quantities, may be good candidates for storage in a **monofill**. Monofills are landfills that take only one type of waste, typically in bales. Monofills are designed to store waste so that it can be easily recovered later. Most waste destined for monofills is high-value waste. Cost-effective recycling methods are now being developed for wastes in monofills. One such type of monofill is for carpet waste; recent advances in separation technology have made carpet monofills into mines for raw materials.

The world's population continues to grow. Given increasing standards of living worldwide, depletion of resources is a concern. Smart use of materials can help ensure an ongoing supply to meet human needs both now and in the future.

2.6.0 Indoor Environment Best Practices

People in developed countries spend over 90 percent of their time (on average) indoors in climate-controlled buildings. These buildings are sealed against leaks to maximize energy efficiency. At the same time, the materials used to build these structures have become largely composite and/or synthetic. From carpets to engineered wood products, much of the indoor environment now consists of products that emit chemicals over their life spans.

Together, these factors have led to a sharp increase in building-related health symptoms known as Sick Building Syndrome. Symptoms include headaches, fatigue, and other problems that increase with continued exposure. With increased concerns for energy conservation, building operators often reduce ventilation rates in unoccupied spaces or during nights and weekends. Reduced ventilation rates, along with tighter building construction and chemical emissions from building products, greatly reduce indoor air quality. In addition, many products such as drywall, carpet, and ceiling tiles serve as excellent food sources for mold. Inadequate ventilation, moisture, and food sources have led to explosions of mold growth in some construction projects. This has had disastrous results for building occupants.

People also contribute to the problem by virtue of their activities. CO_2 levels are an indicator of the effectiveness of ventilation of a room. Higher concentrations of CO_2 correlate with drowsiness, reduced productivity, or even headaches and other health effects. Activities like smoking, cooking, or using printers and photocopiers contribute pollutants to the indoor air. Shedding skin cells and the gases produced from digestion also contribute to poor indoor air quality.

Keeping occupants happy, healthy, and productive is essential for good business and a happy home life. This section explores ways to green buildings by improving indoor environmental quality. One tactic is to prevent potential problems at their source. When pollution sources cannot be avoided, take measures to isolate the source of pollution to prevent it from spreading. Use the laws of physics and psychology to create pleasant and well-functioning spaces for building occupants. Give users control over their environment to allow them to make adjustments for comfort and productivity.

2.6.1 Preventing Problems at the Source

Often, decisions made during the design of a building can have a huge impact on indoor environmental quality. For example, locating buildings upwind of major pollution sources such as power plants can reduce the need to fix air quality later. The same applies to locating buildings away from noise sources such as highways. Careful placement of air intakes away from pollution sources, such as loading docks, can also prevent problems. Be sure all vents and exhaust areas are located downwind from air intakes. Consider the features of neighboring buildings when deciding where to locate air intakes and exhausts.

The materials and processes used to construct a facility have the potential to create problems. Many modern finishes such as paints, sealants, carpets, and composites contain solvents and adhesives that release VOCs as they age. The **off-gassing** of VOCs is an example of **fugitive emissions**. These emissions come from the material itself, not from a stack or vent. That new car smell that many people like is actually caused by the materials releasing VOCs, such as **urea formaldehyde**. If you can smell an odor, the product that caused it is already on its way to your lungs. Select products with a low or zero VOC content. Most major paint producers now have lines of low VOC paints that perform as well and cost about the same as traditional paints. These paints are **water-based** rather than **solvent-based**. Colors may be limited for some brands and types of low VOC paints. The pigments required to achieve extremely dark colors often contain VOCs. Be sure to check the VOC content of pigments when purchasing dark colored paints. Green Seal offers an industry standard for paints to ensure that they meet the criteria for low VOCs and are lead and cadmium free (*Figure 40*). **Green Seal Certified** paints and sealants are listed online at **www.greenseal.org**.

> **WARNING!**
> Just because your project uses low VOC products does not mean you should be careless during installation. Ensure adequate ventilation when installing products that might offgas, and wear appropriate personal protective equipment.

GREEN SEAL LOGO

70101-14_F40.EPS

Figure 40 Look for the Green Seal Certified paint logo.

Other products such as carpets, furnishings, adhesives, and composite wood can also be evaluated using third party standards such as the California Department of Public Health (CDPH) *Standard Method for the Testing and Evaluation of Volatile Organic Chemical Emissions from Indoor Sources Using Environmental Chambers*. Green building rating systems such as the Leadership in Energy & Environmental Design (LEED) rating system reference this and other standards to ensure that products are rigorously and consistently tested for emissions. Many products are tested in enclosed environmental test chambers. VOC emissions from these products are measured over time and reported for use in product labels (*Figure 41*).

Proper design, drainage, and landscaping can help prevent water intrusion into a building. Wherever water or moisture exists at warm temperatures with an available food source, mold or mildew is likely to grow. This can be prevented through good design, ventilation, and maintenance. Make sure the areas surrounding the building are graded properly to drain water away from the building. This helps prevent moisture problems in basements and crawl spaces. Crawl spaces should also have a **waterproof** barrier between the soil and the conditioned space.

Careful selection of landscape plants can also help to prevent problems. Many ornamental trees and shrubs produce a significant amount of pollen. So do species such as olive, acacia, oaks, maples, and pines. Pollen produced by these trees can aggravate allergies and can contaminate air intakes and rainwater harvesting systems. Ask

VOC CONTENT

70101-14_F41.EPS

Figure 41 VOC content can be determined from product labels.

about the pollen production of landscape plants before you make your final selection.

During construction, pay attention to the sequencing of construction activities. Schedule dust-producing and dirty activities before carpets or other absorptive finishes are installed. Other absorptive finishes include wall coverings, fabrics, ceiling tiles, and many types of insulation. Ensure that materials are stored on-site in a way that keeps them safe from moisture and potential mold growth. If possible, use just-in-time delivery to minimize the amount of time materials must be stored on the site. Never install material that has become wet without testing to be sure it will not become contaminated with mold. Wait until after the building shell is complete with windows, doors, and a roof before installing any finishes or materials that can retain moisture. Be sure to allow adequate time for curing of concrete and finishes and provide adequate ventilation. Significant moisture is released during curing and will remain in the building and cause problems if it has no exit.

Select finishes that are nonabsorptive such as tile floors and painted walls (*Figure 42*). These finishes are easier to clean and are water-resistant. They also do not trap particulates and odors that can cause occupant discomfort. Choose finishes that contain biocidal (germ-killing) or mold-resistant coatings. Natural linoleum has biocidal properties, which makes it a good choice for flooring.

After the building is finished, green housekeeping practices can help maintain a healthy indoor environment. Follow manufacturer's instructions for maintaining finishes. For example,

70101-14_F42.EPS

Figure 42 Nonabsorptive finishes such as ceramic tiles are easier to clean and do not absorb pollutants.

vacuum carpets often to prevent the buildup of irritants such as dust mites; read labels on cleaning chemicals, and choose nontoxic, water-based cleaners wherever possible.

Monitor building humidity and maintain relative humidity levels between 30 percent and 50 percent in all occupied spaces of the building. During summer, cooling coils may require maintenance to ensure that they are properly dehumidifying incoming air. Inspect and properly maintain all HVAC equipment; this enables you to identify and fix problems before they create unhealthy conditions.

2.6.2 Providing Segregation and Ventilation

The second step in achieving good indoor environmental quality is to segregate activities and materials that create pollution from other parts of the building. Segregation is especially important during construction activities to prevent contamination of building systems. Use separate ventilation systems for high-risk areas. Entrance control measures such as walk-off mats help block contaminants that can enter the building through entryways.

Any type of construction activity has the potential to cause future indoor air quality problems if not properly managed. Many construction activities, such as cutting operations and sanding, produce dust or airborne particulates. Other activities involve installing materials that release VOCs or moisture while they cure, including paint, sealants, adhesives, concrete, and new lumber. Even with dust collectors installed on equipment, building systems must be protected from absorbing pollutants during construction. These contaminants reduce indoor air quality and may result in damage to sensitive building systems. Proper protective measures reduce cleanup time, which speeds construction and lowers the cost (*Figure 43*).

At a minimum, construction isolation measures should include protection of all HVAC intakes, grills, and registers. Use plastic sheeting and duct tape or ties to seal off all possible points of entry to building ductwork or plenums. Be sure to deactivate the HVAC system before you do this, and isolate entryways to other parts of the building and stairwells or corridors outside the work areas.

Install absorptive materials after dust-producing activities are completed. Absorptive materials include carpet, wall coverings, fabrics, ceiling tiles, and many types of insulation. If this is not possible, use plastic or paper sheeting to protect absorptive surfaces.

Maintain proper ventilation during construction to ensure worker health and safety by exhausting pollutants from the workspace. If possible, maintain continuous negative pressure exhausted to outside air. Exhausting workspace air prevents dust and contaminants from migrating to other parts of the building. Ventilation should be provided by a temporary ventilation system to avoid contaminating building ductwork with pollutants. If the building's permanent ventilation system is used, filters with a minimum efficiency reporting value (MERV) no less than 8 should be used.

Regular housekeeping during construction keeps contaminants under control and prevents them from becoming airborne. Specialty contractors can be retained for this purpose. Alternatively, work practices can be put in place to require workers to keep their areas clean and free of debris. All construction isolation measures should be regularly inspected for leaks and tears and replaced as needed.

At the end of construction, remove all masking and sheeting used to isolate ventilation systems. Temporary high-efficiency filters may be used in the HVAC system during some construction activities and before building occupancy. High-efficiency filters trap remaining contaminants and prevent them from entering the system. Replace all filters prior to occupancy. The EPA recommends a two-week flushout period for all new buildings prior to occupancy. During flushout, ventilation systems are run continuously with full outdoor air, accelerating the offgas removal from building materials. However, it can also introduce large amounts of humidity into the building depending on the climate and season. This humidity adds to the already high humidity rates from the curing of materials inside the building itself. If a building flushout is undertaken, ensure that humidity controls are in place to handle the extra humidity load that may occur.

Research has shown that the primary sources of household pollutants are ordinary products such as paint, cleaning compounds, personal care products, and building materials. Everyday tasks such as bathing, laundering, cooking, and heating can all contribute to poor indoor air quality.

Areas of concern in commercial buildings include kitchens, office equipment rooms, housekeeping areas, and chemical mixing and storage areas. Provide separate ventilation for these areas to minimize the risk of pollutants spreading to other parts of the building. In addition, these areas may be isolated from other spaces in the

70101-14_F43.EPS

Figure 43 Construction isolation measures ensure that contaminants from construction tasks do not contaminate building systems.

building through full deck-to-deck partitions and sealed entryways, depending on the risk level of contamination. Indoor smoking areas are special cases that require careful design and construction to protect nonsmokers from exposure to second-hand smoke.

Entrance control is important to keep dirt and pollutants outside the building from getting inside. Air lock entryways and vestibules are one approach that can also save energy by keeping conditioned air inside the building and unconditioned air out. Entry areas can benefit from the use of walk-off mats and dirt collectors, particularly in areas where inclement weather may lead to snow and ice being tracked into the building. Walk-off mats are also available for use outside construction areas to help prevent contamination of other areas of the building.

2.6.3 Taking Advantage of Natural Forces

A variety of natural forces can affect the indoor environment. The laws of physical science can be used to provide natural lighting, ventilation, heating, and cooling in a space. This not only saves energy, but also can provide better indoor environmental quality for occupants.

Paint colors are one of the easiest ways to influence the experience of a space (*Figure 44*). Often, they require no additional expense. Color can influence both how you perceive spaces and how you feel about them. Dark colors can make workspaces feel cramped and depressing, and tend to make rooms look smaller and more confined. Lighter colors create a sense of openness, may improve mood, and make rooms look larger and more spacious. Dark colors make high ceilings look closer, while lighter colors make low ceilings

70101-14_F44.EPS

Figure 44 Paint color choices can influence the user's experience of a building space.

look higher. Combinations of colors can be especially effective in creating focus areas and drawing people into a space.

Natural daylight can be used to enhance the indoor environment in many ways, but can cause problems if not managed well. Potential problems include excessive heat gain, fading, and glare. Common features to incorporate daylighting in buildings include skylights, atria, and glazing. Light shelves or louvers are one way to incorporate daylight into a space without overheating spaces along the building perimeter. Light shelves bounce sunlight onto the ceiling of a room and reflect it further back into the building space. In combination with reflective or light-colored ceilings, light shelves spread the benefits of natural light throughout a space.

> **WARNING!**
>
> Skylights, atria, and windows offer benefits to building occupants, but they can pose hazards for workers. Energy efficient windows with multiple panes and thermal frames are much heavier than conventional windows. They often come in large sizes in green buildings, and can result in strain injuries during installation. Use proper lifting practices. Cleaning these transparent surfaces often involves work at height. Be sure to employ proper fall protection.

Natural ventilation, when carefully designed, can also create a comfortable indoor environment with minimal energy. *Figure 45* shows an example of a natural ventilation system that would work well in an arid climate. Wind scoops are used to capture prevailing winds and direct them over a receptacle of moisture. As the air absorbs moisture, it becomes cooler and sinks down into the living space. As it mixes in the living space and becomes stale, its temperature gradually increases. The warmer stale air rises to ceiling level and is exhausted downwind with another air scoop. This type of system relies on absorption of moisture to cool the air. In climates with high humidity, other approaches are required.

Finally, geothermal heating and cooling takes advantage of the fact that the earth's temperature below the frost line stays relatively constant year round. Geothermal heating and cooling systems use pipes embedded in the ground or in nearby ponds or streams to shed heat in the summer and absorb heat in the winter. The fluid in these pipes is preheated or precooled by this contact with the earth and used to condition indoor environments. Geothermal heat pumps can also be used

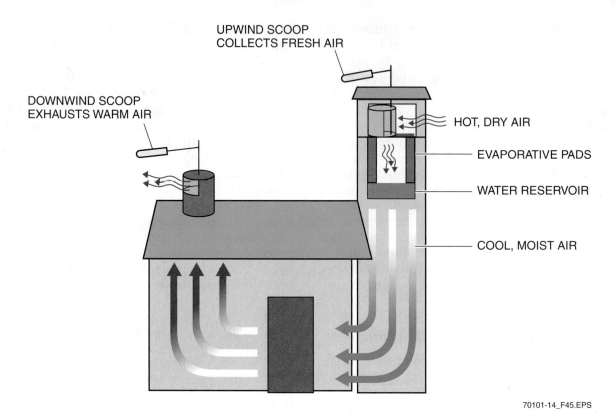

UPWIND SCOOP
COLLECTS FRESH AIR

DOWNWIND SCOOP
EXHAUSTS WARM AIR

HOT, DRY AIR

EVAPORATIVE PADS

WATER RESERVOIR

COOL, MOIST AIR

70101-14_F45.EPS

Figure 45 Natural ventilation system.

to heat hot water. They are very energy efficient, despite a relatively high initial cost.

2.6.4 Giving Users Control over Their Environment

Giving users control over their environment helps to improve indoor environmental quality for building occupants. Various methods can be used to achieve this. Operable windows, climate controls, and underfloor air distribution systems (UFADs) all provide occupants with the ability to control their individual spaces. The effects on building users can be significant. Studies have suggested that occupants who can control their spaces are happier, have greater job satisfaction, take fewer sick days, and are more productive.

Operable windows have gone in and out of vogue in architectural design over the past decades. Traditional buildings without mechanical heating and cooling had operable windows to allow users to control the indoor climate. Many contemporary buildings, however, have mechanical systems that don't work well with uncontrolled introduction of outside air. For this reason, many commercial and institutional buildings do not include operable windows. This creates a threat to passive survivability if the power to run mechanical equipment becomes unavailable. Passive survivability is the ability of a building

to continue to function when the infrastructure goes down. There is growing interest in designing buildings around this idea, as many current buildings operate poorly or fail when deprived of power. In many climates, operable windows can be used in lieu of mechanical ventilation during the swing seasons (fall and spring) when outdoor temperatures are mild. This can save considerable energy. It also contributes to user satisfaction with individual workspaces.

Lighting, temperature, humidity, and ventilation controls are also becoming more sophisticated in modern buildings. Many commercial buildings rely on elaborate sensor and control systems to adjust these variables. However, each individual user of a building is different. Not all users are comfortable with the same environmental conditions. Individual controls at each workstation or office can help users adjust conditions to meet their individual needs (*Figure 46*). It can also increase their satisfaction with the space.

Underfloor air distribution systems are one way to increase the level of control users have over their individual workspaces. These systems provide the ability to control ventilation rates and perceived temperatures at each individual workspace. They also increase ventilation effectiveness and promote a uniform flow of conditioned air through the workspace (*Figure 47*).

The Psychological Impacts of Color

Psychological research has shown that people react to different colors in predictable ways. Common colors and their psychological reactions are as follows:

- Yellow is highly visible and often used for safety markings. Yellow can make small rooms look larger and bring light to narrow entranceways and hallways.
- Orange is perceived to be cheerful and friendly. It is a good choice for rooms where people gather informally.
- Red stimulates the pituitary gland and raises heart rates and blood pressure. It also stimulates appetite. Red encourages action and aggressiveness. It is often used on buttons and knobs.
- Blue is a calming, soothing color. Blue works well in bedrooms.
- Green evokes feelings of relaxation and quietness. Green works well in study or work environments since it promotes concentration.
- Purples can reduce blood pressure and suppress appetite.
- Brown works well in environments where food is prepared or eaten. It also works well in general living environments. Brown is associated with comfort, reliability, and warmth.
- Gray encourages creativity. Gray interiors can be perceived as depressing unless accented with bright, clean colors.
- White is good around food and in precision work environments since it suggests sterility. Pure white can be perceived as harsh. Off-whites are better suited for many applications.
- Black is perceived to be dignified, sophisticated, and elegant. It tends to visually recede and enhances other colors used in combination with it.

These systems can be combined with modular office furniture to allow rapid reconfiguration of space. They make it easy to adjust the configuration of space while maintaining proper ventilation and space conditioning in each work area. Modular control units can be moved around to provide airflow wherever users are located. This level of flexibility and control means that users are happier in their workspace. It also means they are less likely to need supplemental space heaters or fans to be comfortable.

70101-14_F46.EPS

Figure 46 Giving users control over their environment can increase productivity and satisfaction.

2.7.0 Integrated Strategies

Many of the tactics and technologies described in the previous section offer multiple benefits for the green environment. The challenge for you is to find green solutions that meet the owner's needs for a facility. At the same time, they should not damage natural ecosystems or deplete resource bases, and should not cost more than a traditional project would cost. This section highlights strategies for green building decisions that achieve multiple benefits and affect multiple systems. Finding good integrated solutions can help you create better green buildings that cost less to build and have long-term benefits as well.

2.7.1 Solving the Right Problem

The first strategy for greening a project should be to make sure you're solving the right problem. Too often, building professionals start planning a new facility without considering other options. Possibilities include leasing, renovating an existing building, and telecommuting. In some cases, this could result in both green advantages and performance advantages at a much lower cost than building a new structure. In other cases, the most appropriate solution may still be to build new. However, considering other options early in the process can lead to a different mindset about

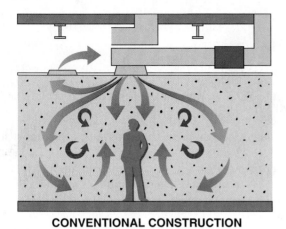

CONVENTIONAL CONSTRUCTION

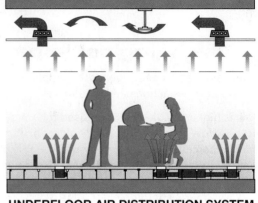

UNDERFLOOR AIR DISTRIBUTION SYSTEM

70101-14_F47.EPS

Figure 47 Underfloor air distribution systems help conditioned air reach all parts of a space.

the project that can help you be more creative on later decisions throughout the project.

The key to making sure you're solving the right problem is to focus not on solutions, but on the needs those solutions will meet. For example, how many times have you been asked at the checkout line whether you'd prefer paper or plastic bags? If you focus on these two choices as the only options available to you, you might miss

other solutions. However, if you phrase the question as, "How can I transport my groceries from store to home?" you can consider other options entirely. What about reusable canvas bags, grocery delivery services, or even modular shopping carts that fold up to fit in your car? You might even decide to grow some of your own food to reduce your need for groceries. To use this principle, determine the functional necessity to be

A goal of some high performance buildings is to be net zero. This means that over the course of a year, the building will produce as much or more of a resource as it consumes. Buildings generally try to be net zero with regard to energy or water. Energy can be produced on site through photovoltaics or other sources, and water can be obtained by rainwater harvesting or water reuse and treatment on site after it is used. The US Army is striving to make its facilities net zero, not just for energy and water but also for solid waste. This includes composting and recycling on site and accepting waste from other facilities. The amount of waste recovered from other sources is used to offset waste from the facility that cannot be recovered on site.

To successfully achieve net zero, a facility must use resources as efficiently as possible. Efficiency improvements are nearly always less expensive than

PHOTOVOLTAICS PRODUCING ENERGY ON A US ARMY FITNESS CENTER IN MONTEREY, CALIFORNIA. (NET ZERO ENERGY SITE)

70101-14_SA19.EPS

on-site generation of power or water. Carefully consider what waste streams are available locally from nearby facilities when planning for net zero waste. On-site waste recovery or recycling can be expensive and take a lot of space; it may not make sense in every situation. However, sharing capacity with neighboring facilities can help make the investment worthwhile.

---

met. If you find yourself defining your problem in terms of one or more solutions, chances are you need to take a step back and reassess your options.

2.7.2 Understanding and Exploiting Relationships Between Systems

Another approach is to exploit the functional relationships between systems. Building systems are related to one another, and the design of one system affects the design of another. For example, increasing the weight of a structure means that the foundation has to be upsized as well. Sometimes these relationships can be exploited to pay for investments in one system through savings in another. This is known as integrative design (*Figure 48*).

For example, consider using high-performance windows to provide daylighting. Considered in isolation, more expensive windows will raise total project cost. However, using high-performance windows will improve the energy efficiency of the building envelope. Additional daylighting will also reduce the heat load from light fixtures. Therefore, the extra costs of better windows can be offset by reducing the capacity of the building cooling system. In addition, a smaller HVAC system might mean smaller pumps, fans, and

motors, reduced duct sizes, smaller plenums, and reduced floor-to-floor height, also reducing the cost of the facility. Reduced floor-to-floor height means less surface area of the building envelope, which means less material costs for the system. It also means that the overall weight of the building is reduced, meaning that foundations can be smaller and more efficient as well.

In the end, the overall increase in the total first cost of the project may be negligible if the benefits of improving one system are captured in the design of related systems. More importantly, life cycle cost savings can be even greater with these more efficiently designed systems. HVAC systems in particular will be much more efficient if they are correctly sized for the facility, allowing them to operate at maximum efficiency over the life cycle of the facility.

2.7.3 Using Services Rather than Products

A third strategy being used to green construction projects is dematerialization. Dematerialization refers to using services instead of products to meet user needs. Construction companies have successfully used this concept for years. For example, a small general contractor may rent equipment that is used infrequently, such as heavy equipment for grading.

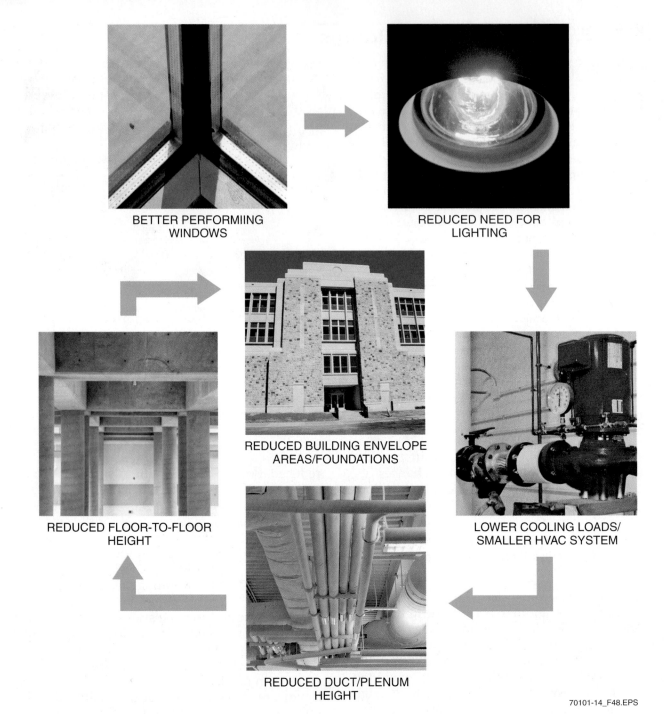

Figure 48 Integrative design means that investments in one system can be offset by savings in related systems.

BETTER PERFORMIING
WINDOWS

REDUCED NEED FOR
LIGHTING

REDUCED BUILDING ENVELOPE
AREAS/FOUNDATIONS

REDUCED FLOOR-TO-FLOOR
HEIGHT

LOWER COOLING LOADS/
SMALLER HVAC SYSTEM

REDUCED DUCT/PLENUM
HEIGHT

70101-14_F48.EPS

Rather than take ownership responsibility (and associated liability) for equipment, you pay another company to provide the benefits of that equipment to you. That company has an incentive to provide the most efficient equipment possible. The company makes its profits based on a fee per hour or unit of service provided. It also has incentive to design products that can be easily repaired, upgraded, or disassembled. This is because the company retains responsibility for the ongoing maintenance and eventual disposition of the equipment.

2.7.4 Considering the Options

When considering what actions to take on a project, it's critical to make smart choices. Every action you take comes with a cost. Recognizing those costs and considering the benefits of your actions can help you choose actions that achieve the results you intend. Think about leverage points; look for easy actions that can make a big difference.

One way to think about potential choices is to consider who has to act. Is action required on the part of building professionals, laypersons, or both? As a building professional, you might not want to change your work practices. However, your knowledge of construction means that you have a much better chance of being successful than a building user getting involved with the operation of the building (*Figure 49*).

Another issue to consider is whether users will have to change their behaviors to achieve the desired effects. Some changes are completely transparent to users, while others require significant change in habits or procedures. Changes are more likely to have the desired outcomes if they do not require people to change their behavior or comfort level. For example, asking occupants to turn down the heat to an uncomfortable level will save energy at first. However, it is a change that is unlikely to be sustained over time (*Figure 50*). When possible, choose solutions that can get the job done without requiring users to change their habits.

The degree to which the change is compatible with existing infrastructure is important. Some changes, like lighting retrofits, can be as simple as changing a light bulb (*Figure 51*). Other changes may not be compatible at all with existing buildings or may not be possible given the skills and equipment of the construction crew. Look for easy changes that can be undertaken with your existing skills and tools wherever possible.

Finally, availability is key to determine how easy a change will be to implement. Imagine if you had to special order every piece of technology you needed to get a job done. It's much easier to make a change when you know you can easily obtain the materials and systems you need to get the job done. Many building material retail stores now stock green products (*Figure 52*). Be sure to ask about these products using the factors discussed in earlier parts of this module.

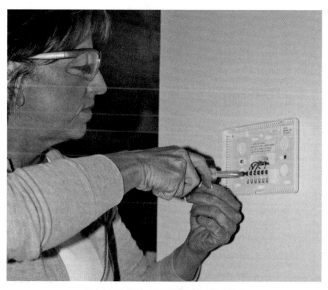

(A) HOMEOWNER CHANGES

(B) DESIGN CHANGES

70101-14_F49.EPS

Figure 49 Changes can be undertaken by professionals, laypersons, or both.

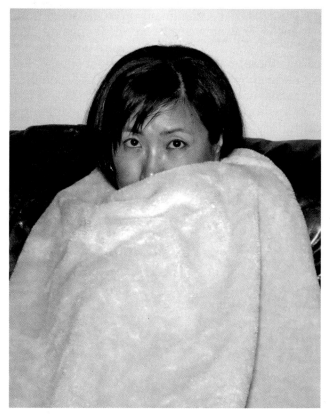

70101-14_F50.EPS

Figure 50 Changes that require users to sacrifice comfort are unlikely to be sustained.

70101-14_F51.EPS

Figure 51 Changes can be simple if they require no modification of infrastructure.

2.7.5 Counting All the Costs

Decisions about individual materials or systems are sometimes made on a first cost basis without considering cost from a life cycle perspective. This means that some green products seem more expensive than they really are. If savings in labor cost and construction schedule were taken into account, these materials could show an immediate cost advantage to owners. For example, one project manager estimated a savings of $3,000 per day due to shortening the construction schedule on a dormitory project. This dormitory was built using pre-engineered concrete panels, enabling the project to be built more quickly. In addition, the owner saved money by avoiding the hotel expenses of dorm residents, who were able to move in sooner.

70101-14_F52.EPS

Figure 52 Some green products are easy to find on the shelves of your local store.

Other factors that can make the cost of green building more reasonable include the following:

- Reduced costs of materials and waste disposal
- Reduced liability and environmental risk
- Happier occupants with increased productivity, reduced absenteeism, fewer building-related health problems, improved morale, and less employee turnover
- Reduced operational and disposal costs
- Reuse of facilities that otherwise would be disposed
- Preparedness for future regulations and requirements

Each of these benefits reflects a potential cost savings for owners. However, many of these types of costs are not typically counted as part of project costs. If these benefits can be realized, then green projects will have an economic advantage over conventional projects. Counting all the costs gives a truer picture of what green best practices will mean for the building over its whole life cycle.

Building Automation

An important trend in green building is building automation. Building automation is an integrated approach to optimize the building's use of water and energy while maintaining occupant comfort. The "nerves" of the system are a series of sensors to measure building conditions. These sensors are connected to a control system, which is the "brain" of the system. The control system uses information from the sensors to adjust different parts of the building to control environmental conditions. For example, daylight harvesting can be used to save energy by using photosensors to turn off lights when adequate daylight is available. The automation system may also activate shades or smart windows to provide shading when there is too much light or heat gain.

Building automation systems can be used to control a wide variety of interior and exterior conditions. These include lighting, temperature, humidity, and others. They may also be used to track performance of building systems and notify operators of potential problems. They also can provide useful information to building users about the effect their actions have on building performance. In this way, building automation can influence both the actions of the building systems and the actions of the building users that affect energy performance.

Did You Know?

Saving Water vs. Saving Energy

Sometimes saving one resource results in unintentionally consuming more of another resource. A recent study at Virginia Tech considered the effects of water recirculation systems, required by code in some areas. These systems prevent wasted water at the tap by keeping hot water circulating so there is no wait for hot water. At the same time, making energy at power plants also consumes water in the form of steam production. The study found that water saved by recirculation devices is less than the water used by power plants to make the energy used to run them. This example shows the need to think about all possible impacts of a choice before deciding what to do.

Additional Resources

Field Guide for Sustainable Construction, Department of Defense – Pentagon Renovation and Construction Office. (2004). PDF, 2.6 MB, 312 pgs. Available for download at www.wbdg.org.

Greening Federal Facilities, 2nd Ed. US Department of Energy Federal Energy Management Program. (2001). PDF, 2.1 MB, 211 pgs. Available for download at www.wbdg.org.

Natural Capitalism. Lovins, A., Hawkin, P., and Lovins, L.H. (1995). Little, Brown, & Company, Boston, MA. Available online at www.natcap.org.

Sustainable Buildings Technical Manual. Public Technologies, Inc./US Department of Energy. (2006). PDF, 3.1 MB, 292 pgs. Available for download at www.greenbiz.com.

Sustainable Buildings and Infrastructure: Paths to the Future. Pearce, A.R., Ahn, Y.H., and Hanmi Global. (2012). Routledge, London, UK.

Sustainable Construction: Green Building Design and Delivery, 3rd Ed. Kibert, C.J. (2012). Wiley, New York, NY.

The HOK Guidebook to Sustainable Design, 3rd Ed. Odell, W. and Lazarus, M.A. (2015). Wiley, New York, NY.

2.0.0 Section Review

1. The percentage of US buildings that suffer from poor indoor air quality is about _____ .

   a. 15 percent
   b. 20 percent
   c. 33 percent
   d. 50 percent

2. The parts of a building site that are paved are called _____ .

   a. softscape
   b. hardscape
   c. landscape
   d. parkscape

3. Most of the water on the planet is _____ .

   a. groundwater
   b. ice
   c. salt water
   d. fresh water

4. A high-albedo roof _____ .

   a. absorbs solar energy
   a. is expensive
   b. reflects solar energy
   c. reduces light pollution

5. Salvaged materials save the use of _____ .

   a. raw materials
   b. labor
   c. equipment
   d. water

6. As they age, modern finishes, like paint, release _____ .

   a. air
   b. VOCs
   c. carbon
   d. energy

7. Changes are easier and more likely to be successful when _____ .

   a. building users are required to change their habits
   b. building occupants agree to turn down thermostats
   c. the construction crew has compatible skills and equipment
   d. suppliers special order the necessary materials

SECTION THREE

3.0.0 CONTRIBUTING TO A PROJECT'S LEED CERTIFICATION

Objective

Explain how craft workers can influence and contribute to a project's Leadership in Energy and Environmental Design (LEED) certification.

a. Describe the LEED rating process.
b. Identify construction activities and project features that affect a project's LEED rating.
c. List kinds of information collected during construction to support LEED documentation.
d. Identify common construction pitfalls that affect a project's LEED rating.

Trade Terms

Corporate Sustainability Report (CSR): An annual report by a company documenting its efforts to reduce negative environmental and social impacts.

EarthCraft: A residential green building program of the Greater Atlanta Home Builders Association in partnership with Southface Energy Institute.

Environmental Product Declaration (EPD): A document listing the environmental impacts caused by manufacture of a product.

Halons: Ozone-depleting compounds consisting of bromine, fluorine, and carbon. They were commonly used as fire extinguishing agents, both in built-in systems and in handheld portable fire extinguishers. Halon production in the United States ended on December 31, 1993.

Heat island: An area that is warmer than surrounding areas due to absorbing more solar radiation.

Hydrochlorofluorocarbons (HCFCs): A group of man-made compounds containing hydrogen, chlorine, fluorine, and carbon. They are used for refrigeration, aerosol propellants, foam manufacture, and air conditioning. They are broken down in the lowest part of the atmosphere and pose a smaller risk to the ozone layer than other types of refrigerants.

Leadership in Energy and Environmental Design (LEED): A point-based rating system used to evaluate the environmental performance of buildings. Developed by the US Green Building Council, LEED provides a suite of standards for environmentally sustainable construction. Founded in 1998, LEED focuses on the certification of commercial and residential buildings and the neighborhoods in which they exist.

Location valuation: A bonus under the LEED rating system for using a product produced locally. Materials from within a 100-mile radius can be counted as twice their value as a reward for using local products.

Persistent, bioaccumulative toxin (PBT): A harmful substance such as a pesticide or organic chemical that does not quickly degrade in the natural environment but does accumulate in plants and animals in the food chain.

Sedimentation: The process of depositing a solid material from a state of suspension in a fluid, usually air or water.

Spoil pile: Excavated soil that has been moved and temporarily stored during construction.

Thermal bridging: A condition created when a thermally conductive material bypasses an insulation system, allowing the rapid flow of heat from one side of a building wall to the other. Metal components, including metal studs, nails, and window frames, are common culprits.

Waste separation: Sorting waste by specific type of material and storing it in different containers to facilitate recycling.

Choosing the best green practices to use on a project can be difficult, but there are tools to help. Green building rating systems help you evaluate how buildings affect the green environment. Rating systems are particularly important to owners and purchasers of buildings. They offer valuable information about the building's overall environmental performance. Green building rating systems provide a way to help the construction market do better in terms of meeting green project goals. Most green building rating systems include performance levels that must be met for the building to be certified. They may also provide guidelines that help project teams meet or exceed those levels.

There are two major types of green building rating systems: location-specific and general. Location-specific rating systems are typically developed by local builder associations. They take into account local climate and building best practices for the area. Most location-specific rating systems apply to residential buildings. There are over 25 different local or regional residential rating systems in the United States, the first of which was established in Austin, Texas in 1991. Other local rating systems include EarthCraft homes in the southeastern United States, the Built Green rating system in Colorado (*Figure 53*), and the California Green Builder program for production builders in California.

The second type of rating system is general, and these systems can be used in multiple locations. In the United States, there are both residential and commercial general rating systems. There are also rating systems for neighborhoods and for infrastructure such as roads. Many national rating systems used in other countries are modeled after the US Green Building Council's Leadership in Energy and Environmental Design (LEED) rating system (see **www.usgbc.org**), developed in the 1990s. The LEED rating system was based on a similar system called the Building Research Establishment Environmental Assessment Method

(BREEAM), which was developed earlier in the United Kingdom.

National residential rating systems include LEED for Homes and the National Association of Home Builders' National Green Building Standard (**www.nahbrc.org**). Commercial rating systems in use in the United States include the LEED rating system, Living Building Challenge (**living-future.org/lbc**), and the Green Globes rating system (**www.thegbi.org** and **www.greenglobes.com**).

The LEED Green Building Rating System is the predominant standard in the United States for rating commercial buildings (*Figure 54*). LEED is a reference standard for government agency buildings at the federal, state, and local levels. Many owners in the private sector have also adopted it. The US General Services Administration requires all of its new buildings to meet at least the Gold level of LEED certification. Other federal agencies have followed suit. Many states also require LEED certification for public buildings at various levels. Even individual cities have adopted LEED as a standard, including the cities of New York, Atlanta, Seattle, and others. By 2015, McGraw Hill Construction estimates that approximately 40 to 48 percent of all new non-residential construction projects in the United States will be green.

3.1.0 LEED Green Building Rating System

The LEED rating system applies to a wide variety of project types. It is designed to be applicable in different climates and contexts throughout the United States and worldwide. According to the

70101-14_F53.EPS

Figure 53 The Built Green program certified this home outside Denver, Colorado.

70101-14_F54.EPS

Figure 54 The LEED Green Building Rating System ensures that buildings are constructed to specific environmental standards.

Rating Systems Around the World

Green building rating systems were first developed in the United Kingdom in the 1990s. Since then, they have spread around the world. A recent study by the National Institutes of Occupational Safety and Health found over 40 different rating systems worldwide. Rating systems have been used on all continents except Antarctica. Certification by a rating system is required in some locations for some types of projects. However, in most places, certifying a project is voluntary. Project teams can also use rating systems as an internal tool to improve their projects without certification. Many tools are available free online.

US Green Building Council, by the end of 2013, approximately 42 percent of all square footage of buildings pursuing LEED Certification was outside the United States. LEED is characterized in terms of its structure, seven major categories of credits, and four levels of certification. A complete list of the LEED prerequisites, credits, and point values for new construction (LEED-BD+C Version 4) is provided in *Appendix A*.

> **NOTE**
>
> This module focuses primarily on the LEED for Building Design + Construction (LEED-BD+C) rating system, in Version 4 at the time of this writing. Other rating systems differ by project type and may vary in specific credit requirements.

3.1.1 Structure of the LEED Rating System

LEED for New Construction, now known as LEED for Building Design + Construction (LEED-BD+C), was the first of all the LEED rating systems to be developed. Each of the subsequent systems was modeled after the same structure.

The LEED rating system consists of a series of performance goals and requirements in nine primary categories:

- *Integrative Process (IP)* – This category addresses the process by which the project is planned, designed, and delivered.
- *Location and Transportation (LT)* – This category covers issues related to the location of the project site and the amenities around and near the project that affect transportation needs.
- *Sustainable Sites (SS)* – This category covers impacts to the project site during construction and the impacts the project will have on neighboring sites after completion.
- *Water Efficiency (WE)* – This category addresses water use and wastewater generation by the building during operation.
- *Energy and Atmosphere (EA)* – This category covers all aspects of the building's energy performance, energy source(s), and atmospheric impacts.
- *Materials and Resources (MR)* – This category pertains to the sources and types of materials used on the project, the amount of waste generated, and the degree to which the project makes use of existing buildings.

GOING GREEN

Seattle Focuses on Saving Power

Seattle, Washington, has written LEED certification into its City Plan. In addition to providing city staff to assist builders in creating and building green projects, they have legislated that all city-funded projects and renovations with over 5,000 square feet of occupied space must achieve a LEED Gold rating.

70101-14_SA21.EPS

- *Indoor Environmental Quality (EQ)* – This category covers aspects of the building's indoor environment, ranging from ventilation to air quality to daylight and views.
- *Innovation in Design (ID)* – This category rewards the project for going beyond the minimum credit requirements and for using a LEED Accredited Professional.
- *Regional Priority (RP)* – This category rewards projects for addressing local environmental priorities (identified by ZIP code at **www.usgbc.org**).

Each category consists of a series of credits that define points that can be earned by a project. Six out of the nine categories also have prerequisites that the project must meet to be considered for certification. A project must meet all prerequisites in all categories in order to pursue certification. Projects can receive credit for features of special regional importance with Regional Priority (RP) credits. They can also receive credit for going significantly beyond basic credit requirements with Innovation in Design (ID) credits. For example, a system that pays special attention to water efficiency in a desert environment might be given an extra RP credit in addition to a WE credit.

3.1.2 Types of Rating Systems and Levels of Certification

The LEED system was initially developed to apply to new commercial construction. As the system grew in popularity, it became apparent that different types of projects would require different criteria to be properly rated. The original rating system was then customized through the development of special credits for specific project types such as hospitals and schools. These types of projects have characteristics that require special interpretation of LEED credit requirements. However, they can still use the basic LEED-BD+C structure. Separate versions of the rating system exist in the following major categories:

- LEED for Building Design and Construction (BD+C), which has special credits applicable to core and shell, schools, retail, data centers, warehouses and distribution centers, hospitality, healthcare, homes and multi-family lowrise, and multi-family midrise projects.
- LEED for Interior Design and Construction (ID+C), which addresses commercial interiors as well as special requirements of retail and hospitality projects.
- LEED for Building Operations and Maintenance (O+M), which covers existing buildings and has special credits for retail, schools, hospitality, data centers, and warehouses and distribution centers.
- LEED for Neighborhood Development (ND), which can be applied to developments in the planning phase or at the completion of the development.

Some parts of different rating systems complement each other. For instance, a LEED-BD+C project can receive up to fifteen points for choosing a project site within a development certified under LEED-ND. Likewise, a LEED-ID+C project can receive credit for being located in a building certified under LEED-BD+C.

There are 110 possible points under the LEED-BD+C rating system, and twelve prerequisites that apply to all projects. Projects must meet all prerequisites applicable to the project to pursue certification. Certification can be achieved at four different levels:

- *Certified* – 40 to 49 points
- *Silver* – 50 to 59 points
- *Gold* – 60 to 79 points
- *Platinum* – 80 points and above

Platinum is the most difficult level to reach. Currently, just over 4,000 certified projects have achieved this level of certification, out of more than 70,000 worldwide. This is only about 6 percent of all certified projects. All LEED rating systems award certification at the four listed levels.

3.1.3 Certification Process

In 2008, the United States Green Building Council established the Green Building Certification Institute (GBCI) to administer LEED project certifications and professional credentials and certificates. The process of certifying a building under LEED has several steps. These steps are guided by the LEED Online documentation system. Certification occurs in parallel with an integrative process that involves participation of many stakeholders. The first step in the plan is to undertake discovery and set a direction for the project. Based on the type of project, the appropriate rating system is chosen. Then, Minimum Program Requirements are verified. The team will review project information and hold a goal-setting workshop including the owner, design team, and construction team. This should all occur before design begins. A formal LEED project scope will be defined, including a project boundary.

Formal certification begins by registering the project with GBCI to declare intent to pursue certification. This allows the project team to access GBCI databases as well as set up an online workspace to manage project documentation. Registering the project requires paying a fee to GBCI that is the same for all projects.

After the project has been registered, the next step is to identify LEED credits that can be pursued. This is based on the project goals, and typically involves a meeting to review the LEED checklist (*Figure 55A and B*). The process of documentation to prove compliance with LEED credit requirements continues throughout construction. It may also extend into the first year of occupancy depending on the credits pursued.

The project team can submit documentation to GBCI for review at two points in time. The first time is at the end of the design process, and the second is after construction is complete. The project team may also opt to submit everything at once when the project is complete. All documentation is compiled by the project team using LEED Online. Submittal of documentation for review involves paying a review fee based on project size and the rating system used. This prompts GBCI to conduct the review. If the team submits a design review, GBCI will evaluate points under the rating system that can be measured at the end of the design process. It will not evaluate credits that require documentation during the construction phase. Design phase review is useful for the project team to get an idea of how many points they are likely to obtain in the project. It provides a basis for deciding how many additional points will need to be earned during construction to meet the project goals.

At the conclusion of the project, final documentation is assembled online. A fee is paid, and the package is reviewed by GBCI. Upon review of the full project documentation, GBCI may elect to request additional clarification as part of a point audit. GBCI makes a final determination as to which points should be awarded and determines a level of certification for the project. The project team has the right to appeal any credits declined in the review. This requires paying an additional fee and providing additional documentation. The ruling of GBCI following appeals is final.

3.2.0 Goals of the LEED Green Building Rating System

The basic concepts of the LEED rating system can be captured in the form of eight simple goals:

1. Choose a good site for your project.
2. Protect and restore the surrounding environment.
3. Promote sustainable behavior.
4. Conserve resources.
5. Find better waste sinks.
6. Create healthy and productive living environments.
7. Check systems to be sure they work right.
8. Look for better ways to do things.

> **NOTE**
>
> The details of LEED credit requirements are beyond the scope of this module. If you're interested in learning more about specific LEED credit requirements, you can view LEED standards for any of the current rating systems at **www.usgbc.org**.

Credit	Integrative Process	1

Location and Transportation		**16**
Credit	LEED for Neighborhood Development Location	16
Credit	Sensitive Land Protection	1
Credit	High Priority Site	2
Credit	Surrounding Density and Diverse Uses	5
Credit	Access to Quality Transit	5
Credit	Bicycle Facilities	1
Credit	Reduced Parking Footprint	1
Credit	Green Vehicles	1

Sustainable Sites		**10**
Prereq	Construction Activity Pollution Prevention	Required
Credit	Site Assessment	1
Credit	Site Development - Protect or Restore Habitat	2
Credit	Open Space	1
Credit	Rainwater Management	3
Credit	Heat Island Reduction	2
Credit	Light Pollution Reduction	1

Water Efficiency		**11**
Prereq	Outdoor Water Use Reduction	Required
Prereq	Indoor Water Use Reduction	Required
Prereq	Building-Level Water Metering	Required
Credit	Outdoor Water Use Reduction	2
Credit	Indoor Water Use Reduction	6
Credit	Cooling Tower Water Use	2
Credit	Water Metering	1

Energy and Atmosphere		**33**
Prereq	Fundamental Commissioning and Verification	Required
Prereq	Minimum Energy Performance	Required
Prereq	Building-Level Energy Metering	Required
Prereq	Fundamental Refrigerant Management	Required
Credit	Enhanced Commissioning	6
Credit	Optimize Energy Performance	18
Credit	Advanced Energy Metering	1
Credit	Demand Response	2
Credit	Renewable Energy Production	3
Credit	Enhanced Refrigerant Management	1
Credit	Green Power and Carbon Offsets	2

70101-14_F55A.EPS

Figure 55A LEED project worksheet (1 of 2).

Materials and Resources		13
Prereq	Storage and Collection of Recyclables	Required
Prereq	Construction and Demolition Waste Management Planning	Required
Credit	Building Life-Cycle Impact Reduction	5
Credit	Building Product Disclosure and Optimization - Environmental Product Declarations	2
Credit	Building Product Disclosure and Optimization - Sourcing of Raw Materials	2
Credit	Building Product Disclosure and Optimization - Material Ingredients	2
Credit	Construction and Demolition Waste Management	2

Indoor Environmental Quality		16
Prereq	Minimum Indoor Air Quality Performance	Required
Prereq	Environmental Tobacco Smoke Control	Required
Credit	Enhanced Indoor Air Quality Strategies	2
Credit	Low-Emitting Materials	3
Credit	Construction Indoor Air Quality Management Plan	1
Credit	Indoor Air Quality Assessment	2
Credit	Thermal Comfort	1
Credit	Interior Lighting	2
Credit	Daylight	3
Credit	Quality Views	1
Credit	Acoustic Performance	1

Innovation		6
Credit	Innovation	5
Credit	LEED Accredited Professional	1

Regional Priority		4
Credit	Regional Priority: Specific Credit	1
Credit	Regional Priority: Specific Credit	1
Credit	Regional Priority: Specific Credit	1
Credit	Regional Priority: Specific Credit	1

TOTAL Possible Points:		**110**

Figure 55B LEED project worksheet (2 of 2).

3.2.1 Selecting the Site

The first goal of the LEED rating system is to help you choose a good site for your project. One way to get points is to choose a project site in a neighborhood certified under LEED-ND (LT Credit: LEED for Neighborhood Development Location). LEED rewards projects that avoid sensitive land, which includes wetlands, habitats of threatened or endangered species, or flood plains (LT Credit: Sensitive Land Protection). The rating system also awards credits for choosing a site in areas that are already developed to a certain density (LT Credit: Surrounding Density and Diverse Uses), as shown in *Figure 56*. These sites are preferred over undeveloped sites that can require the use of cars to get to amenities such as stores, parks, banks, and others, as shown in *Figure 57*. Locating near transit can also obtain a point under LT Credit: Access to Quality Transit. LEED also encourages projects to locate on sites that are brownfields or located in priority development areas such as historic districts (SS Credit: High-Priority Site). Brownfields are sites that have real or perceived environmental contamination, such as old gas stations, industrial sites, and dry cleaners.

3.2.2 Protecting and Restoring the Surrounding Environment

The second major goal of the LEED system is avoiding or repairing damage to the site, building, surroundings, and world at large during and following construction. Toward this end, project teams can get credit for assessing the ecological resources available on site (SS Credit: Site Assessment). They also set credit for designing

Figure 56 Sites close to amenities reduce the need to travel by car.

70101-14_F57.EPS

Figure 57 Projects in undeveloped areas force people to use motorized transport.

projects to preserve or restore site ecology (SS Credit: Site Development – Protect or Restore Habitat and SS Credit: Open Space).

The construction team plays a critical role in achieving the credits and prerequisite under this goal. A prerequisite for all LEED projects is to meet all requirements of the 2003 US EPA Construction General Permit, which includes measures to protect soil and water streams, and to mitigate dust and noise (SS Prerequisite: Construction Activity Pollution Prevention). All projects over one acre in size must comply with these requirements under federal law. This prerequisite requires projects to

---

Did You Know?

Accreditation vs. Certification

A common mistake when discussing LEED projects is to confuse accreditation and certification. Accreditation is a process that is applied to building professionals who manage the process of LEED certification for projects. Becoming a LEED Accredited Professional requires you to pass the LEED Accreditation Exam (more information is available at **www.gbci.org,** the website of the Green Building Certification Institute that manages and administers the LEED exam). Certification, on the other hand, applies to buildings. Buildings are certified under the LEED rating system and are awarded certification at the Certified, Silver, Gold, or Platinum levels. Keeping these terms straight will help your credibility with respect to the LEED rating system. Remember, people are accredited and buildings are certified.

---

use erosion controls to prevent stormwater run-off, sedimentation, and dust from leaving the site (*Figure 58*). Another consideration is managing excavated soil (*Figure 59*). The EPA General Permit Requirements include stabilizing spoil piles.

LEED encourages builders to limit site disturbance during construction to a minimal area (SS Credit: Site Development – Protect or Restore Habitat), using techniques such as fencing shown in *Figure 60*. LEED also encourages ecological site restoration at the end of construction. You can help meet this requirement by being aware of how far you can go outside your work boundary. If you are working on a LEED project, ask your supervisor to mark the boundaries and be sure you stay within them.

Contractors should install protective fencing when working near trees. Place the fencing at or beyond the drip line of the tree's outermost branches to protect the root system (*Figure 61*). Notify your supervisor if you see unprotected trees.

Since parking lots are significant sources of heat and pollutants in the built environment, LEED rewards projects that minimize the area of the site used for parking (LT Credit: Reduced Parking Footprint). Parking is also an important consideration in SS Credit: Rainwater Management and SS Credit: Heat Island Reduction. These credits focus on reducing specific negative impacts of pavements used for parking. Treating and/or retaining rainwater on site using other means is also encouraged (SS Credit: Rainwater Management). This helps prevent contamination of local waterways by runoff from parking lots and roofs.

70101-14_F58.EPS

Figure 58 Sedimentation fencing in place to manage stormwater runoff.

FENCING TO LIMIT DISTURBED AREA

70101-14_F59.EPS

Figure 59 Grass seeding and straw mulch are one way to stabilize excavated soil.

TREE PROTECTION FENCING

70101-14_F60.EPS

Figure 60 Temporary fencing can be used to mark limits of disturbance.

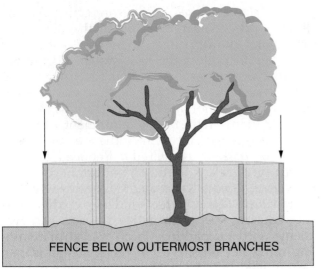

Figure 61 Effective tree protection fencing requires the tree to be protected as far out as the drip line.

Projects that use light-colored roofs are also recognized for reducing urban heat islands (SS Credit: Heat Island Reduction). Urban heat islands are developed areas where ambient temperatures are higher than surrounding areas. They are caused by dark-colored pavements and buildings that absorb solar energy (*Figure 62*).

LEED also acknowledges projects that avoid using unnecessary levels of lighting (SS Credit: Light Pollution Reduction) due to unshaded light fixtures, such as the one shown in *Figure 63*. Light pollution disturbs endangered species such as loggerhead turtles. It is suspected of disrupting sleep and mating cycles in animals, and may have adverse effects on people.

To protect the environment outside the job site, projects are encouraged to reduce or eliminate the use of ozone-depleting chemicals. This includes chemicals in air conditioning and refrigeration equipment and propellants in insulation (EA Prerequisite/Credit: Fundamental/Enhanced Refrigerant Management). Chlorofluorocarbons (CFCs), hydrochlorofluorocarbons (HCFCs), and Halons all deplete atmospheric ozone. Damaging atmospheric ozone has negative effects on plants, animals, and people. Producing energy from renewable sources instead of fossil fuel is also encouraged under LEED. This helps reduce the amount of carbon released into the atmosphere contributing to the greenhouse effect. EA Credit: Renewable Energy Production and EA Credit: Green Power and Carbon Offsets both encourage renewable energy. Renewable energy can be produced on-site or can be purchased from off-site utilities.

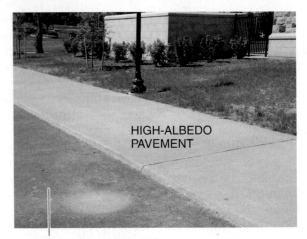

Figure 62 Dark pavements worsen urban heat islands.

Figure 63 Many outdoor fixtures contribute to light pollution.

Several LEED credits recognize the impacts that building materials have on the natural environment. LEED rewards projects that use materials where manufacturers openly disclose these impacts (MR Credit: Building Product Disclosure and Optimization – Environmental Product Declarations). Project teams can receive additional points if the products they choose are environmentally preferable. One way to do this is if the raw materials used to make those products have been developed responsibly (MR Credit: Building Product Disclosure and Optimization – Sourcing of Raw Materials). Another way is to reuse buildings or building materials to prevent the need to dispose of them and make new materials (MR Credit: Building Life-Cycle Impact Reduction). Project teams can also use techniques such as life cycle assessment to measure impacts of their choices in building design.

For Materials and Resources credits, it is very important to track products used in construction throughout the construction process. Be sure to ask your supervisor what criteria are important for different building products. Be sure to verify any substitutions before you make them. Keep track of documentation for products you use, since it will be needed to document the certification. For instance, MR Credit: Building Product Disclosure and Optimization – Material Ingredients requires information from manufacturers about the ingredients used to make materials. MR Credit: Building Product Disclosure and Optimization – Sourcing of Raw Materials requires information about how those ingredients were extracted from the natural environment and what effects were caused.

3.2.3 Promoting Sustainable Behavior

The next major goal of the LEED rating system encourages projects to include amenities that promote sustainable behavior. This includes providing amenities that encourage people to use low-impact transportation such as transit, bicycling, green vehicles, or carpooling (LT Credits: Access to Quality Transit; Bicycle Facilities; Green Vehicles), as shown in *Figure 64*. LT Credit: Surrounding Density and Diverse Uses provides the opportunity to walk or bicycle to nearby amenities instead of driving. SS Credit: Open Space rewards projects that provide green space, which may encourage building occupants to spend time being active outdoors.

Water and energy performance credits require the building to include efficient fixtures that save water and energy during its life cycle (WE Prerequisite/Credit: Indoor Water Use Reduction

70101-14_F64.EPS

Figure 64 Bike racks, reserved parking for carpools, and preferred parking for motorcycles and high-efficiency vehicles reward green behavior.

and EA Prerequisite/Credit: Minimum Energy Performance; Optimize Energy Performance).

Projects that include water or energy meters can also help building occupants behave more sustainably. WE Prerequisite: Building-Level Water Metering and EA Prerequisite: Building-Level Energy Metering require all LEED projects to have at least one meter for the whole building's water and energy use. WE Credit: Water Metering and EA Credit: Advanced Energy Metering offer additional points for measuring water or energy use for individual systems or devices.

Recycling facilities (*Figure 65*) encourage occupants to sort their solid waste and reduce the amount of trash that goes to landfills (MR Prerequisite: Storage and Collection of Recyclables). Having facilities to collect and store recyclables is required of all LEED certified buildings.

Providing smoking facilities as required by EQ Prerequisite: Environmental Tobacco Smoke (ETS) Control helps protect nonsmokers from environmental tobacco smoke. When adequate smoking facilities are not provided, smokers tend to huddle near the door, which may cause discomfort for those entering and leaving the building. Many hospitals, schools, and government buildings prohibit smoking within a certain distance of building entrances.

3.2.4 Conserving Resources

The fourth goal encompassed by LEED is to conserve resources as much as possible throughout the project. This includes using environmentally preferable materials wherever possible. Environmentally preferable materials include products whose manufacturers have used best practices for reducing impacts during manufacture, or whose products are extracted responsibly (MR Credit: Building Product Disclosure and Optimization – Sourcing of Raw Materials). Products made from raw materials such as FSC Certified wood, rapidly renewable materials, or with recycled content can be counted in this credit. LEED also rewards manufacturers who make information about their product available to help designers make better choices (MR Credit: Building Product Disclosure and Optimization – Environmental Product Declaration and MR Credit: Building Product Disclosure and Optimization – Material Ingredients). This includes information about a product's environmental impacts and ingredients. Finally, this goal also includes using less material, or reusing existing materials instead of new materials (MR Credit: Building Life-Cycle Impact Reduction).

Buildings are also rewarded by LEED for using graywater or rainwater instead of potable water for landscaping (WE Prerequisite/Credit: Outdoor Water Use Reduction). LEED also rewards indoor uses of graywater such as toilet flushing (WE Prerequisite/Credit: Indoor Water Use Reduction), such as the under-sink system shown in *Figure 66*. Conserving water or finding alternate sources of water for cooling towers is rewarded under MR Credit: Cooling Tower Water Use.

Finding better sources for electrical energy is encouraged via installation of on-site power generation (EA Credit: Renewable Energy Production). This helps a building be more energy independent from outside fuel sources such as foreign oil. You can also get credit for contracting for green power from a third party provider or for purchasing carbon offsets (EA Credit: Green Power and Carbon Offsets). Including capabilities for peak shaving and controlling a building's energy demand is also rewarded (EA Credit: Demand Response). Using natural daylighting (EA Credit: Daylighting) also conserves energy that would be used to artificially light a building.

Not only is the type of material important for a green building, but also where the components come from. This helps reduce impacts associated

70101-14_F65.EPS

Figure 65 Recycling facilities encourage users to sort their waste.

70101-14_F66.EPS

Figure 66 An under-sink graywater system stores sink and shower water for toilet flushing.

with transportation of materials. Several LEED MR credits reward the use of regional materials to reduce transportation impacts. *Figure 67* shows an example of the 100-mile radius that defines a regional material according to LEED. Products whose extraction, manufacture, and purchase occurs within 100 miles of the project site can be counted double in LEED calculations.

As a craft worker, you may be required to procure green materials and systems to meet the project specifications (*Figure 68*). When procuring these materials, carefully review the specifications to be sure you understand all product requirements. Keep cut sheets or other documentation to support compliance. These materials will help when preparing the project documentation.

Better sources for construction materials that are encouraged under LEED include the following:

- Salvaged or reused materials
- Materials with recycled content, especially post-consumer recycled content
- Materials from locations near the project site
- Rapidly renewable bio-based materials
- Wood products that are sustainably harvested
- Materials from a manufacturer who will take back or recycle the product at the end of its life

A. Paper fiber panel core
B. Laminated strand lumber (LSL)
C. Engineered wood I-joist
D. Abundant American hardwoods
E. Laminated veneer lumber (LVL)
F. Adobe block with straw
G. Straw-core SIP with oriented strand board (OSB) skin
H. Parallel strand lumber (PSL)

70101-14_F68.EPS

Figure 68 Bio-based materials are an important resource for green building.

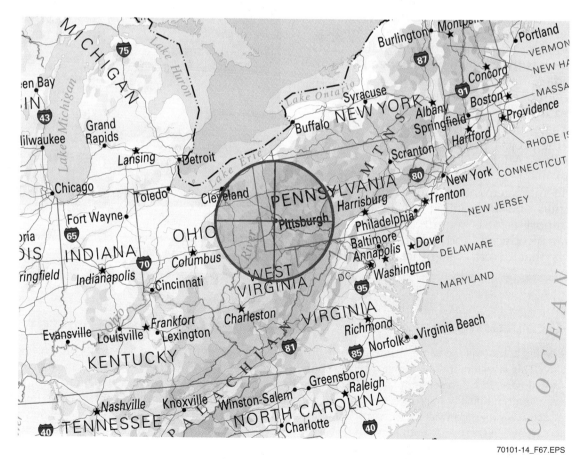

70101-14_F67.EPS

Figure 67 A regional material originates from within 100 miles of a job site.

Manufacturers must provide corporate sustainability reports (CSR), environmental product declarations (EPD), or other objective documentation to prove their claims about their products. A corporate sustainability report provides information about a company's programs and achievements each year to benefit the environment and society. An environmental product declaration is an objective report that documents the environmental impacts of a product and allows it to be compared to other products. Some of the impacts in an EPD include global warming potential, ozone depletion, depletion of nonrenewable resources, and acidification of water sources. Products whose EPDs are produced by the manufacturer instead of a third party receive only half value in LEED calculations. Third party-verified EPDs are more objective, since the third party has no motive to rate the product better than it actually is.

3.2.5 Finding Better Waste Sinks

Minimize or prevent waste wherever possible using the best practices discussed earlier in this module. During design and construction, the focus of LEED is on material use and waste. LEED offers credit for designing buildings to use fewer materials under MR Credit: Building Life-Cycle Impact Reduction. Healthcare projects also can get special credits for designing spaces for flexibility (MR Credit: Design for Flexibility). This is because healthcare facilities change often to accommodate new practices. Each change can produce a lot of solid waste unless spaces are designed to allow easy changes.

With waste that is produced, be sure you understand the project policy for waste separation and recycling. All LEED projects must develop and implement a construction and demolition waste management plan (MR Prerequisite: Construction and Demolition Waste Planning). A final report is also required to show what waste was produced and what happened to it. Projects that divert a significant amount of construction and demolition waste from landfills or incineration can get additional points (MR Credit: Construction and Demolition Waste Management). Construction projects can meet this credit requirement by sending mixed waste to an off-site separation facility (*Figure 69*) or through on-site separation of construction and demolition waste for recycling (*Figure 70*).

Projects can also achieve credits for reducing the amount of waste generated during construction to less than 2.5 pounds of construction waste per square foot (12.2 kilograms of waste per square meter). This goal can be achieved in various ways discussed in Section Two, including the following:

- Prefabrication of components
- Modular construction
- Designs that use standard-size materials

If on-site recycling separation is part of the project, be sure to put waste in the correct dumpster (*Figure 71*). If you don't see a place to recycle, ask for one. Do not contaminate dumpsters with other waste. In particular, keep treated wood

70101-14_F69.EPS

Figure 69 In some areas, mixed construction waste can be taken off site and sorted for recycling.

Did You Know?
Pre-Consumer vs. Post-Consumer Recycled Content

Not all recycled content is the same. Some waste is produced as a byproduct of manufacturing, then recovered for recycling. This is called pre-consumer recycled content because it is recycled before it enters the market to be used by consumers. It is also sometimes called post-industrial recycled content. Often, the manufacturing process could be made more efficient to reduce or eliminate this waste. Post-consumer recycled content is the result of a consumer actively recycling waste after a product has been used. One example is recycled plastic bottles being made into carpet. LEED rewards both types of recycled content, but pre-consumer content can only be counted half as valuable as post-consumer content in LEED calculations.

70101-14_F70.EPS

Figure 70 In other areas, on-site sorting is required for recycling.

70101-14_F71.EPS

Figure 71 On-site waste separation can earn up to two LEED credits plus one Innovation Credit.

separate from other materials. Also, keep anything with food on it out of cardboard dumpsters. Cleaner dumpster loads are more likely to be accepted for recycling.

If you have excess materials or salvaged goods on a project, ask your supervisor to arrange a place for temporary storage. Consider donating unused materials to a Habitat for Humanity ReStore. If there is a ReStore near you, take the time to visit it; you may find products that you want to buy.

LEED also includes credits to encourage finding better waste sinks during building operation. SS Credit: Rainwater Harvesting encourages behavior to prevent stormwater from being treated as waste. Various Water Efficiency credits also reduce the amount of wastewater that is produced by the building. MR Prerequisite: Storage and

Collection of Recyclables provides user amenities to divert solid waste from being sent to landfills.

3.2.6 Creating Healthy and Productive Living Environments

The sixth major goal of the LEED rating system is to create healthy and productive living environments for building occupants. LEED rewards projects for encouraging human-powered transport. This can be achieved by locating within walking distance from other amenities (LT Credit: Surrounding Density and Diverse Uses). Projects can also provide bicycle facilities (LT Credit: Bicycle Facilities) and outdoor open space (SS Credit: Open Space) to be used by occupants.

At the building scale, LEED strives to make buildings healthy and productive by the following actions:

- Avoiding toxic ingredients when choosing building materials (MR Credit: Building Product Disclosure and Optimization – Material Ingredients).
- Achieving minimum levels of indoor air quality (EQ Prerequisite: Minimum Indoor Air Quality Performance).
- Keeping nonsmokers away from environmental tobacco smoke (EQ Prerequisite: Environmental Tobacco Smoke [ETS] Control)
- Making sure spaces receive adequate ventilation (EQ Credit: Enhanced Indoor Air Quality Strategies).
- Making sure air quality is good before allowing occupants to move in (EQ Credit: Indoor Air Quality Assessment).
- Avoiding the use of materials that emit high levels of VOCs (EQ Credit: Low-Emitting Materials). *Figure 72* shows a product label indicating VOC content.
- Separating areas of the building that contain polluting activities from other areas (EQ Credit: Enhanced Indoor Air Quality Strategies).
- Designing building conditioning systems to provide thermal comfort (EQ Credit: Thermal Comfort).
- Providing users with the ability to control conditions in their own spaces (EQ Credit: Thermal Comfort and EQ Credit: Interior Lighting).
- Providing users with access to natural daylight and views (EQ Credit: Daylight and EQ Credit: Quality Views).
- Enhancing well-being and productivity through controlling acoustic conditions (EQ Credit: Acoustic Performance).

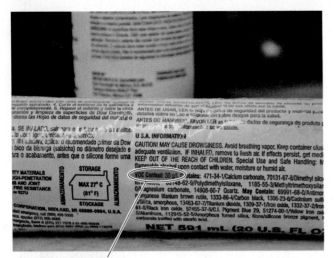

VOC CONTENT

70101-14_F72.EPS

Figure 72 Look for levels of VOCs on product labels and buy products with low levels.

If you work inside the building, make sure you understand how your activities affect indoor air quality. LEED provides a credit for creating and following an indoor air quality management plan during construction. This is based on SMACNA's *Indoor Air Quality Guidelines for Occupied Buildings Under Construction* and other measures, including the following:

- Prohibiting the use of tobacco products on site during construction. This includes smoking and smokeless tobacco inside the building. It also forbids use of tobacco products within 25 feet (7.5 meters) of the building's entrances.
- Protecting absorptive materials from moisture damage before, during, and after installation.
- Keeping contaminants out of all HVAC equipment. This includes not operating building air handling equipment during construction unless it is properly protected with filters. It also includes sealing ductwork during construction.
- Avoiding toxic materials where possible.
- Isolating areas of work to prevent contamination of other spaces.
- Scheduling construction activities to finish dirt-producing activities before installing absorptive materials.
- Maintaining good job site housekeeping.

Managing moisture during construction is key for the health of building occupants. Prevent moisture damage by storing materials in a protected staging area. Never install wet materials. Installing wet or damp materials can create mold problems in the building before it is even complete. If materials do get wet, they may need to be replaced.

Consider whether your activities will produce airborne dust or offgas VOCs. Take measures to protect the building from these activities (*Figure 73*).

To create buildings that are free of problems for future occupants, focus on work inside the building. Using low-emitting materials can help to ensure good indoor air quality (*Figure 74*).

LEED encourages separate ventilation for chemical mixing and storage areas and copy rooms (*Figure 75*). These areas can produce air pollution that is unhealthy for building occupants. Copiers and printers produce ozone during operation, which can cause respiratory irritation.

70101-14_F73.EPS

Figure 73 Containing construction contaminants is critical to ensure indoor air quality.

LOW/ZERO-VOC PAINT, STAINS, AND SEALANTS   LOW-VOC ADHESIVES (COVE BASE, CARPET)

LOW-VOC CARPET TILES

UREA FORMALDEHYDE-FREE COMPOSITE WOOD

70101-14_F74.EPS

Figure 74 Low-emitting materials can help to ensure good indoor air quality.

Figure 75 Photocopiers are a common source of indoor air pollution in office environments.

Other building products can also be unhealthy. Healthcare projects receive special consideration under LEED for this goal. Persistent, bioaccumulative, and toxic (PBT) materials should be avoided. These materials degrade very slowly and can accumulate in body tissues. They can cause problems with the central nervous system. Special actions that apply to Healthcare projects include the following:

- Reducing the use of mercury-containing products and devices, and preventing mercury release (MR Prerequisite/Credit: PBT Source Reduction - Mercury).
- Reducing the use of lead, cadmium, and copper (MR Credit: PBT Source Reduction – Lead, Cadmium, and Copper).
- Choosing furnishings that don't contain toxic materials (MR Credit: Furniture and Medical Furnishings).

Mercury is a common material used in fluorescent lamps, mercury vapor high-intensity discharge lamps, and some thermostats and switches. These products should be carefully handled if they are used in the project, and avoided if possible. Lead is often found in solder and flux for plumbing connections. It may also be used in roofing and flashing, wiring, and paints. Cadmium is used in paints. Copper can be released into the water supply when copper pipes corrode. All joints in copper piping should be

either mechanically crimped or properly soldered to prevent corrosion.

Furnishings may also contain toxic materials, and LEED encourages projects to use furnishings that do not contain them. Toxins in furniture can come from stain and non-stick treatments, antimicrobial treatments, heavy metals, and urea formaldehyde. Furnishings also can offgas VOCs, which may be released in large quantities into a new building when furnishings are installed. Consider unpacking furnishings off-site to allow them to offgas before moving them into the building.

LEED credits to promote daylighting, acoustic performance, and quality views are included because of growing evidence that these factors provide many benefits to occupants (*Figure 76*). Exposure to natural daylight has been correlated with effects ranging from better test scores to improved dental health. Being able to look out the window has also been linked with increased productivity in offices and faster healing in hospitals. LEED encourages the use of daylighting to increase human comfort as well as reduce the need to rely on electric lighting. Better acoustics enhances learning, and has been made a prerequisite in LEED for Schools projects.

3.2.7 *Performing System Checks*

LEED puts special emphasis on checking the design, construction, and initial operation of a building to be sure it meets the design intent. This requires a process called commissioning (EA Prerequisite: Fundamental Commissioning and Verification). Commissioning can also be expanded to cover more systems at a more detailed level (EA Credit: Enhanced Commissioning).

Figure 76 Daylight and views are correlated with many positive outcomes.

Commissioning involves a third party in design review, testing and balancing of systems, and system turnover (*Figure 77*).

At the prerequisite level, mechanical, electrical, plumbing, and renewable energy systems must be commissioned to obtain LEED certification. Commissioning agents will evaluate these systems for energy and water consumption, indoor environmental quality, and durability. They will also review the building's exterior enclosure design to ensure it has been properly designed and constructed. Projects can receive additional points for expanding commissioning activities to include in-depth design review, training, and post-construction verification.

LEED also rewards projects that include the ability to measure the performance of systems. This includes water consumption (WE Prerequisite: Building-Level Water Metering and WE Credit: Water Metering) and energy consumption (EA Prerequisite: Building-Level Energy Metering and EA Credit: Advanced Energy Metering). These systems may also be able to adjust the building's mechanical and electrical systems to optimize performance. Another example is using CO_2 monitors to control building ventilation (EQ Credit: Enhanced Indoor Air Quality Strategies). The building itself should also be designed to keep occupants comfortable (EQ Credit: Thermal Comfort and EQ Credit: Acoustic Performance). LEED also encourages post-occupancy evaluation after the building has been occupied. The purpose of this survey is to ensure that building users are still comfortable after the building has been broken in.

70101-14_F77.EPS

Figure 77 Building commissioning provides a chance to check that all building systems work well together.

3.2.8 Seeking Better Methods

An important part of green building is improving the process by which projects are delivered. LEED requires an integrative planning and design process that involves stakeholders outside the normal process (IP Prerequisite: Integrative Project Planning and Design). Designers from multiple disciplines along with owners, constructors, and operators participate throughout the planning and design phases. The combination of perspectives, along with an open mind for new ideas, can lead to design breakthroughs. It also helps to ensure that what is designed can actually be built and operated properly. Teams that go beyond the basic process can earn additional points (IP Credit: Integrative Process). This

GOING GREEN

High-Performance Building Envelopes

Rinker Hall on the University of Florida's Gainesville campus earned LEED Gold certification for its energy-efficient design. The architect was challenged to design a building that would fit in with the other brick buildings on campus, but not trap heat. The compromise was a freestanding masonry shade wall that serves as a second skin and helps match the design of the building to the site. The high-performance building envelope reflects heat and includes an air-infiltration barrier, a thermal break, and high-performance glass.

70101-14_SA22.EPS

includes analyses to identify interrelationships among building systems.

LEED also rewards projects for finding new ways to make projects green that are not currently included in the rating system (ID Credit: Innovation). Project teams can propose new credits for future versions of LEED based on best practices they have achieved. Or, they can pursue a credit from the LEED Pilot Credit Library. The Pilot Credit Library contains credits still under development that have been field-tested by at least one project. However, these credits are not yet a part of the official rating system standard. One example of a current pilot credit is Clean Construction. This credit reduces negative impacts from equipment idling and exhaust from construction vehicles and equipment (*Figure 78*). Projects can also receive innovation credits for greatly exceeding regular credit thresholds in areas including water efficiency, energy performance, materials use, and waste diversion. These are called exemplary performance credits. Finally, projects can be awarded a point for involving a LEED Accredited Professional (LEED AP) as part of the project team (ID Credit: LEED Accredited Professional).

Be aware of the project's goals for LEED certification and speak up if you have ideas for improvement. On the job, various Innovation Credits such as Clean Construction may impact you. Another possible impact may happen if your project is pursuing Innovation Credits for education. You may see students on the job site who are observing or recording your work as part of LEED documentation of the project. Or you may be asked to participate in special training yourself. Feel free to ask questions. If you see students

70101-14_F78.EPS

Figure 78 Using next-generation emissions controls for clean construction, such as the Tier IV engine on this grader, can help earn a LEED innovation Credit.

observing, it is important to show them your typical work practices so they can accurately capture what happens on the job site.

3.3.0 LEED Documentation

Managing documentation for a LEED project can be challenging. The LEED AP handling the documentation of your project will be grateful if your materials and activities are well documented. Be sure you understand the types of information you will be required to provide, and ask questions if needed.

Projects pursuing Materials and Resources credits require information about each product used in the project. This varies based on what specific credits are being pursued. All calculations are based on the material cost of each product, so the actual cost of each product must be provided or estimated. This does not include labor or other installation costs.

For recycled content, you need the percent by weight of both post-consumer and pre-consumer recycled content for all parts of the material. As you might imagine, this can get quite complicated for a product such as carpeting, which includes several different materials. Product documentation may also require information about VOC content, Forest Stewardship Council chain of custody, or other information.

Another challenging factor to calculate is a product's location valuation. This calculation requires that you know the origin of the material and its components. A material may be counted double for many Materials and Resources credits if it originated and was processed and sold within a 100-mile radius of the site.

For Construction and Demolition Waste Recycling, documentation is needed about the total weight or volume of waste for the project, along with the weight or volume of material diverted for recycling. Your project team will choose either weight or volume as the basis for calculations. After the choice has been made, all calculations must use that basis. You should retain waste hauler receipts showing the destination and weight or volume of each different material claimed for this credit. Receipts are also required for products donated to charity. Materials reused on site or reused on other projects require a recorded estimated weight or volume. Photographs of recycling containers are also required.

Photographs are an important part of LEED documentation. They are required as proof of what was done during construction, since many measures will not be visible after the fact.

There is a limited window of opportunity to take required photos. It is critical that photos be taken at the right times for LEED credits being pursued. Construction indoor air quality measures such as HVAC protection are one type of photograph required. Another type is photographs of recycling containers for construction and demolition waste management. Stormwater protection measures such as sedimentation fencing must also be photographed. Be sure to ask your supervisor what photographs you are required to take for LEED documentation. Show the date the picture was taken in the photograph if possible, and take photos on a regular basis.

3.4.0 Common Pitfalls During Construction

With all the ways you can contribute to a LEED project, there are also plenty of pitfalls to avoid. Four of the most serious pitfalls include poor planning, poor execution of credit requirements, poor documentation, and lack of coordination with other trades.

3.4.1 Poor Planning

Some LEED pitfalls occur before the project starts. Contractors must be familiar with all LEED requirements for the project, and they must make all subcontractors aware of those requirements when planning the job. Often, LEED requirements are included as a separate section in the specifications. However, many subcontractors only read the specs pertaining to their trade. This means they can miss LEED requirements that would affect their bid. Contractors must ensure all subcontractors understand LEED requirements when bidding on a project. Ideally, LEED requirements will be included in the

specifications in each section where they apply. Including contractors in integrative design can help to avoid these pitfalls.

It is also important to keep LEED goals in mind when planning project tasks. The sequence of tasks in the schedule is important. Absorptive materials should be properly protected until it is safe to install them. Concrete and other materials should be cured before moisture-sensitive products are installed. Dirt-producing activities should be completed before finishes are installed.

Following a site plan is also important, as construction activities can be damaging to the natural environment. Limits of disturbance should be marked off before clearing begins. LEED projects may require extra dumpsters to account for recycling. Dumpsters should be located to minimize traffic. Material storage and staging is also important; just-in-time delivery can help avoid disturbance to the site.

Make sure that all products used in the project meet LEED requirements. For instance, it is easy to remember that paint should be low- or zero-VOC. However, fire caulking and duct sealants must also be low-VOC to meet air quality limits. Subcontractors should plan to be sure these materials are available. They may need to allow extra time for delivery for some products.

Coordination with the owner and other stakeholders is key. Air quality testing can be disturbed if the owner moves in before testing is done. The owner's furniture may offgas VOCs into the space, leading to bad test results. Housekeeping at the end of a project can also cause problems. If toxic chemicals are used to clean the building, air quality can suffer. This happens even though the materials in the building are low VOC.

Roles in managing LEED requirements are important. Someone must be assigned to verify all products that come on site to be sure they comply. Someone should review all product substitution requests to make sure new products meet LEED goals. Someone must be in charge of collecting documentation. This includes FSC chain of custody information, photographs, and waste hauling receipts. That person should be asked if documentation is complete before final payments are made.

3.4.2 Poor Execution of Credit Requirements

Even if you know exactly what is necessary to achieve a LEED credit, it is not always easy to meet those requirements. Some credits can be difficult to achieve. For example, EQ Credit: Construction Indoor Air Quality Management Plan requires you to protect the indoor air quality of the building. This requires various measures to isolate your work area and keep it clean. Failure to maintain these measures, even if you set them up correctly, can cause the project's goals not to be met.

A project can fail its indoor air quality test for many reasons. This means it would not be awarded the associated LEED credit. Common reasons projects fail IAQ tests include the following:

- Not controlling products brought on site for use in the building
- Allowing high VOC products to be used for fire caulking, duct sealing, and other applications not specifically mentioned by LEED
- Allowing owners to occupy the building before IAQ testing, including moving furniture into the building
- Using toxic materials for final cleaning of the project. These include toxic chemicals and strippers, window washing chemicals, and spot removers, especially spot cleaners for vinyl tile

It is important to make sure everyone involved in the project knows about LEED goals and what must be done to meet them. Even the cleaners brought in at the end of the project are an important part of the project team.

Another potential problem area is SS Prerequisite: Construction Activity Pollution Prevention. As a prerequisite, these requirements must be met. If they are not, the project's LEED certification can be denied. One example of a pollution prevention activity is sedimentation and erosion control. This includes measures like sedimentation fencing and protection around stormwater drains. If you are responsible for maintaining these controls, be sure you understand exactly what needs to be done. Check these systems regularly and repair them if they become damaged. Even if they are not directly your responsibility, inform your supervisor of any problems so they can be corrected.

Similarly, be sure you understand project goals for SS Credit: Site Development – Protect or Restore Habitat. Often areas of the site will be selected to keep from damage during construction. Protection fences may be put up to remind equipment operators to avoid these areas. Ask your supervisor if you are unclear on what can and cannot happen in those areas. Bring problems to the attention of your supervisor or crew leader.

MR Credit: Construction and Demolition Waste Management is also a common credit where things go wrong. Be sure you understand the project's recycling goals and how they will be met. Ask questions in weekly or daily toolbox meetings, especially if you notice that others are not recycling properly. If you see contamination in any of the recycling bins, mention it to your supervisor. Talk to your crewmates to be sure they understand the importance of correct separation for recycling.

Buying green materials for a project can also be a challenge. The specifications for a green project have been carefully developed to meet project green goals. They should be followed whenever you buy materials. Ask vendors for documentation to support the products you buy. Make sure you allow enough time to purchase any unusual materials.

3.4.3 Poor Documentation

The third major pitfall that can impede your project's LEED certification is poor documentation. The amount of information needed to certify a project is extensive; you must know the properties of the products you use, and you also must report their cost. Therefore, keeping track of information as it occurs can help you successfully provide the information the project team needs.

One credit that requires careful documentation is MR Credit: Construction and Demolition Waste Management. For this credit, you must prove where all of the waste for your project has gone. This is necessary to calculate the percent of waste that was diverted from landfills. To do this, you must save waste hauling receipts and receipts from recycling or salvage companies. Set up a process for tracking this information.

What's Wrong with This Picture?

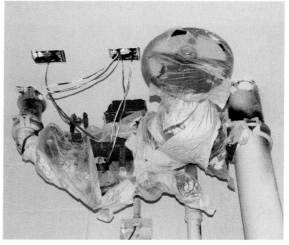

70101-14_SA23.EPS

Another common error in waste documentation is to vary the units used to measure waste. You can choose whether to measure waste by weight (pounds or tons; kilograms) or volume (cubic yards or cubic meters). However, you must choose one way or the other, then use only that basis for all documentation. Be sure to ask about what units will be used by your waste hauling provider. Make them aware of the project's LEED goals.

Photos are also required for many LEED credits. Often, these photos must be taken at a specific time to capture a construction activity. Make sure those times are noted in the schedule. Someone must have responsibility to take photos at the right time. All photos should be properly stored and labeled. Photos are required for EQ Credit: Construction Indoor Air Quality Management Plan, SS Prerequisite: Construction Activity Pollution Prevention, and MR: Construction and Demolition Waste Management.

Another credit requiring product information is EQ Credit: Low-Emitting Materials. This credit requires you to show that carpets, paints, sealants, adhesives, composite wood, and other products meet VOC emission requirements. The easiest way to do this is to ensure that each product meets requirements before you bring it on the job site. After they are on site, keep an inventory of products you have used and their labeled levels of VOCs. For carpets and paints that are certified via a label, be sure to note the label when checking for compliance (*Figure 79*).

The same approach applies for Materials and Resources Credits, which deal with various green materials. For each product used, retain a cut sheet or other data showing relevant properties

VOC CONTENT

70101-14_F79.EPS

Figure 79 Verify that products meet specifications by checking VOC levels on labels.

for each credit. Keep track of the costs of each product so that the LEED AP can track it as part of LEED calculations.

One product that requires special attention is FSC certified wood. To be counted for the credit, the complete chain of custody for that product must be documented, from the initial harvest through to the final supplier. If the FSC chain of custody documentation is not available, it may be possible to ask the supplier to provide additional documentation to fix the problem. However, this must be done as soon as the product is received. If you wait until the end of the project to deal with this problem, you may lose the credit because you can no longer get documentation.

It is important to track documentation as you go, so errors or missing information can be addressed immediately. Do not wait to organize documentation until the end of the project. The person responsible for LEED documentation should be consulted before all final payments are approved. A final review by USGBC should be complete before final payment, ensuring that contractors address any remaining questions for certification.

3.4.4 Lack of Coordination with Other Trades

The fourth major pitfall to achieving LEED certification is a lack of coordination with other trades. Poor coordination may put LEED certification at risk, resulting in poor energy performance and other problems.

Energy performance typically suffers the worst from poor coordination. Cracks and gaps between materials installed by different trades can

What's Wrong with This Picture?

70101-14_SA24.EPS

allow conditioned air to leak from the building, causing significant energy loss or moisture problems. Careful construction follow-up can seal these leaks, though this is often overlooked.

Coordination problems can also contribute to **thermal bridging**. Thermal bridging occurs where materials of higher conductivity (like metal and wood) pass completely through an exterior wall without a thermal break (*Figure 80*). These materials act as thermal pipelines. They can result in cold spots in the wall or ceiling surface where moisture can condense. Over time, these moist areas serve as a breeding ground for mold.

Lack of coordination can also affect structural integrity. Conflicts inevitably occur when multiple trades must fit their systems into the same space. In wood frame construction, improper notching can occur when plumbing or electrical trades install piping or conduit through wood members. The *International Building Code* limits the depth and location of cuts in all structural members (*Figure 81*). This is to ensure that structural integrity is preserved.

All field modifications should be checked to be sure they do not impact the building's integrity. Careful coordination of the trades in these areas prevents the need for rework. Rework can be difficult and expensive after finishes are complete. Coordination can also reduce the risk of failure.

Another common problem is disturbing insulation in walls. The energy performance of insulation is significantly reduced when it is compressed or when gaps are left in wall cavities. It may also be compromised if the insulation gets wet. Check all insulation before interior finishes are installed, allowing it to be fixed while the walls are still open.

During the project, speak up if you see anything that might compromise the project's LEED goals. Make your supervisor aware if you see problems with stormwater management, dust containment, or any other protective measures. If you see others doing things incorrectly, talk with them to find out why. Encourage them to follow the LEED requirements to meet the project goals.

Everyone involved in the project should be made aware of LEED goals. They should know not only the goals that affect them, but all the goals in general. Sometimes trades are familiar with one aspect of green building but not others. Mistakes can be made that can cause the project to lose points. For instance, HVAC workers are familiar with energy performance: they know to seal ductwork to enhance the energy performance of their systems. However, they may not realize that duct sealant contains VOCs. They

70101-14_F80.EPS

Figure 80 Thermal bridging caused by a metal beam.

may use materials that offgas and cause the project to fail its IAQ test. All stakeholders need to understand how their work influences the project's green goals, both directly and indirectly.

Projects seeking LEED certification can perform better than conventional buildings. However, achieving LEED certification requires careful attention during construction. To successfully reach LEED goals, make sure LEED requirements are in the specs and that the project plan meets the requirements. Be sure that what gets done on the project matches the project plan. Together, these actions will help a project be successfully certified.

What's Wrong with This Picture?

70101-14_SA25.EPS

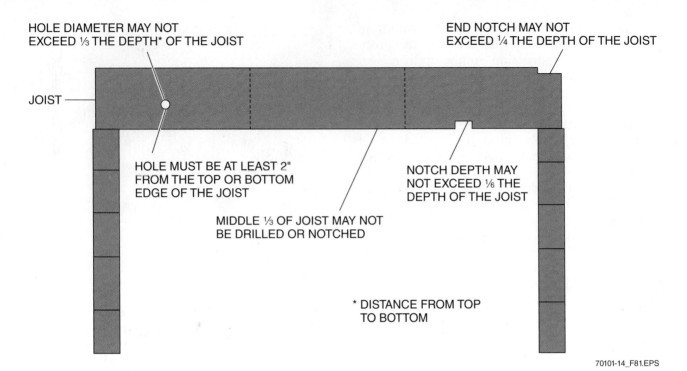

HOLE DIAMETER MAY NOT
EXCEED ⅓ THE DEPTH* OF THE JOIST

END NOTCH MAY NOT
EXCEED ¼ THE DEPTH OF THE JOIST

JOIST

HOLE MUST BE AT LEAST 2"
FROM THE TOP OR BOTTOM
EDGE OF THE JOIST

NOTCH DEPTH MAY
NOT EXCEED ⅙ THE
DEPTH OF THE JOIST

MIDDLE ⅓ OF JOIST MAY NOT
BE DRILLED OR NOTCHED

* DISTANCE FROM TOP
TO BOTTOM

70101-14_F81.EPS

Figure 81 Be careful when notching structural members to preserve structural integrity.

Think About It

Your Responsibilities

Perhaps the most important question you can ask is whether your project is seeking a LEED certification or not. If the project is seeking certification, the next step is to ask what your role will be in getting the certification. If you are procuring materials, what information is needed and who should receive it? What specifications and requirements do the materials need to meet? If you are working outside the building or inside in a protected area, what do you need to do to protect the work area? How should waste be managed? Are there any other special requirements that will be your responsibility? Do you see any opportunities for improvement?

Additional Resources

Field Guide for Sustainable Construction. Department of Defense – Pentagon Renovation and Construction Office. (2004). PDF, 2.6 MB, 312 pgs. Available for download at **www.wbdg.org**.

Greening Federal Facilities, 2nd Ed. US Department of Energy Federal Energy Management Program. (2001). PDF, 2.1 MB, 211 pgs. Available for download at **www.wbdg.org**.

Natural Capitalism. Lovins, A., Hawkin, P., and Lovins, L.H. (1995). Little, Brown, & Company, Boston, MA. Available online at **www.natcap.org**.

Sustainable Buildings Technical Manual. Public Technologies, Inc./US Department of Energy. (2006). PDF, 3.1 MB, 292 pgs. Available for download at **www.greenbiz.com**.

Sustainable Buildings and Infrastructure: Paths to the Future. Pearce, A.R., Ahn, Y.H., and Hanmi Global. (2012). Routledge, London, UK.

Sustainable Construction: Green Building Design and Delivery, 3rd Ed. Kibert, C.J. (2012). Wiley, New York, NY.

The HOK Guidebook to Sustainable Design, 3rd Ed. Odell, W. and Lazarus, M.A. (2015). Wiley, New York, NY.

3.0.0 Section Review

1. The LEED rating system was developed by the _____ .
   a. Lean Building Institute
   b. US Green Building Council
   c. Sustainable Built Environment Initiative
   d. United Kingdom

2. Erosion controls used to prevent stormwater runoff during construction are rewarded by _____ .
   a. LT Credit: Sensitive Land Protection
   b. SS Prerequisite: Construction Activity Pollution Prevention
   c. SS Credit: Rainwater Management
   d. LT Credit: High Priority Site

3. Which credit requires proof of disposal, such as waste hauling receipts?
   a. MR Prerequisite: Storage and Collection of Recyclables.
   b. MR Prerequisite: Construction and Demolition Waste Management Planning.
   c. MR Credit: Building Life-Cycle Impact Reduction.
   d. MR Credit: Construction and Demolition Waste Management.

4. A common pitfall to LEED certification during construction is _____ .
   a. lack of coordination with other trades
   b. using diesel engines on the job site
   c. not following directions
   d. using nonrenewable materials

SUMMARY

As the green environment changes, the construction industry must also change. Past practices have had many negative impacts on the green environment. Today, there are many things that can be done to change for the better.

You can estimate your individual impact by calculating your carbon footprint. Your carbon footprint measures how much carbon your activities emit to the atmosphere. Some of the ways to reduce your carbon footprint include reducing product and energy use, or recycling waste. You can also plant vegetation, find better sources of energy, or buy carbon offsets to help other people improve. As you consider your options, be sure to look for leverage points: small actions you can take to make a big difference.

As a craft worker, there are six major areas where you can improve a construction project's impact on the green environment. These include the site and landscape, water and wastewater, energy, materials and waste, indoor environment, and integrated strategies. Within each of these areas, you can have the most impact by eliminating unnecessary uses of resources. You can also find more efficient ways to use resources. Look for better sources for resources, and better sinks for waste. Seek out integrated strategies where one action has multiple benefits. Again, look for leverage points, where small actions can make a big difference.

The LEED Green Building Rating System provides a useful measuring stick for industry improvement. LEED has nine categories of credits (Integrative Design, Location and Transportation, Sustainable Sites, Water Efficiency, Energy and Atmosphere, Materials and Resources, Indoor Environmental Quality, Innovation in Design, and Regional Priority). Projects can be certified at four levels: Certified, Silver, Gold, and Platinum. Platinum is the most difficult rating level. LEED projects must meet certain basic requirements called prerequisites, and then can choose from points in the nine categories to achieve a rating level.

As a craft worker, you have many opportunities to contribute to a project's LEED certification. How you manage the project site is important. Pay attention to stormwater management and control systems. Be aware of how you protect building systems from potential indoor environmental contamination. Be careful to buy materials that meet specifications. Look for better ways to dispose of waste streams. Remember, your actions can make a real difference.

1. A natural change in the green environment is _____ .
   a. deforestation
   b. urban sprawl
   c. changes in the seasons
   d. aquifer depletion

2. The overuse of groundwater is leading to _____ .
   a. loss of biodiversity
   b. global climate change
   c. aquifer depletion
   d. deeper coal mines

3. One of the contributors to global warming is _____ .
   a. aquifer depletion
   b. deforestation
   c. rising sea levels
   d. recycling

4. Airplanes are among the largest contributors to _____ .
   a. helium emissions
   b. nitrogen emissions
   c. carbon emissions
   d. oxygen emissions

5. The longest phase of the building life cycle is typically _____ .
   a. planning
   b. design
   c. construction
   d. operation and maintenance

6. A way to minimize the urban heat island effect is to _____ .
   a. pave with asphalt
   b. choose light-colored pavements
   c. choose dark-colored pavements
   d. coat paved areas with a sealer

7. The first step in greening water use is to _____ .
   a. identify alternative toilets
   b. eliminate unnecessary uses
   c. install low-flow toilets
   d. install a water meter

8. A common technique to reduce the actual flow of water in faucets is _____ .
   a. turning off the water
   b. aeration
   c. heating
   d. cooling

9. A green alternative source of water is _____ .
   a. groundwater
   b. rainwater harvesting
   c. melting ice
   d. salt water

10. Lighting retrofits to a building have a _____ .
    a. slow payback period
    b. rapid payback period
    c. large payback period
    d. negative payback period

11. Green power generates energy from _____ .
    a. nonrenewable resources
    b. coal
    c. renewable resources
    d. natural gas

12. Material that is recycled from manufacturing processes is termed _____ .
    a. post-consumer waste
    b. consumer waste
    c. post-industrial waste
    d. pre-industrial waste

13. An example of a multi-function material is _____ .
    a. structural insulated panels
    b. hardwood flooring
    c. concrete
    d. steel framing

14. A smart material changes in response to _____ .
    a. funding
    b. the user's needs
    c. environmental conditions
    d. a schedule

15. Rapidly renewable materials are _____ .
    a. recycled
    b. bio-based
    c. unsustainable
    d. expensive

16. Construction isolation measures to prevent building contamination should protect _____ .
    a. storage areas
    b. HVAC intakes
    c. water supplies
    d. construction equipment

17. Passive survivability is the ability for a building to function when _____ .
    a. the infrastructure goes down
    b. there is a labor strike
    c. there is a shortage of supplies
    d. the air is contaminated

18. The LEED rating system that is for the individual tenant spaces in commercial buildings is _____ .
    a. LEED-EB
    b. LEED-NC
    c. LEED-CI
    d. LEED-CS

19. A project team can submit documentation for LEED credits _____ .
    a. only at the beginning of construction
    b. only at the end of construction
    c. only at the end of design
    d. both at the end of construction and design

20. A prerequisite for all LEED projects is to meet all the requirements of the _____ .
    a. Lean Construction Institute
    b. EPA General Construction Permit
    c. LEED requirements
    d. local building code

Trade Terms Introduced in This Module

Absorptive finish: A surface finish that will absorb dust, particles, fumes, and sound.

Acidification: A process that converts air pollution into acid substances, leading to acid rain. Acid rain is best known for the damage it causes to forests and lakes. Also refers to the outflow of acidic water from metal and coal mines.

Aerated autoclaved concrete (AAC): A lightweight, precast building material that provides structural strength, insulation, and fire resistance. Typical products include blocks, wall panels, and floor panels.

Aeration: Mixing air into a liquid substance.

Albedo: The extent to which an object reflects light from the sun. It is a ratio with values from 0 to 1. A value of 0 is dark (low albedo). A value of 1 is light (high albedo).

Alternative fuel: Any material or substance that can be used as a fuel other than conventional fossil fuels. They produce less pollution than fossil fuels. These include biodiesel and ethanol.

Aquifer: An underground layer of water-bearing rock or soil from which groundwater can be extracted using a well.

Aquifer depletion: A situation where water is withdrawn from its underground source faster than the rate of natural recharge.

Best Management Practice (BMP): A way to accomplish something with the least amount of effort to achieve the best results. This is based on repeatable procedures that have proven themselves over time for large numbers of people.

Bio-based: A material derived from living matter, either plant or animal.

Biodegradable: Organic material such as plant and animal matter and other substances originating from living organisms. These are capable of being broken down into innocuous products by the action of microorganisms.

Biodiversity: A measure of the variety among organisms present in different ecosystems.

Biofuel: A solid, liquid, or gas fuel consisting of or derived from recently dead biological material. The most common source is plants.

Biomimicry: The act of imitating nature to create new solutions modeled after natural systems.

Bioswale: An engineered depression designed to accept and channel stormwater runoff. A bioswale uses natural methods to filter stormwater, such as vegetation and soil.

Blackout: A situation where the electrical grid fails and does not provide any power.

Blackwater: Water or sewage that contains fecal matter or sources of pathogens.

Brownfield: Property that contains the presence or potential presence of a hazardous substance, pollutant, or contaminant. Cleaning up and reinvesting in these properties reduces development pressures on undeveloped, open land and improves and protects the environment.

Brownout: A situation where the electrical grid provides less power than normal, but enough for some equipment to still work.

Building-integrated photovoltaics (BIPVs): Photovoltaic systems built into other types of building materials. See *photovoltaic*.

Byproducts: Another product derived from a manufacturing process or a chemical reaction. It is not the primary product or service being produced.

Carbon cycle: The movement of carbon between the biosphere, atmosphere, oceans, and geosphere of the Earth. It is a biogeochemical cycle. In the cycle, there are sinks, or stores, of carbon. There are also processes by which the various sinks exchange carbon.

Carbon footprint: A measure of impact human activities have on the environment. It is determined by the amount of greenhouse gases produced. It is measured in pounds or kilograms of carbon dioxide.

Carbon neutral: An action or product whose production absorbs as much carbon from the atmosphere as it produces.

Carbon offset: An agreement with another party that they will reduce their carbon production by some amount in exchange for payment.

Carpool: An arrangement in which several people travel together in one vehicle. The people take turns driving and share in the cost.

Charrette:: An intense meeting of project participants that can quickly generate design solutions by integrating the abilities and interests of a diverse group of people.

Chlorofluorocarbons (CFCs): A set of chemical compounds that deplete ozone. They are widely used as solvents, coolants, and propellants in aerosols. These chemicals are the main cause of ozone depletion in the stratosphere.

Cob construction: An ancient building method using hand-formed lumps of earth mixed with sand and straw.

Commissioning: A review process conducted by a third party that involves a detailed design review, testing and balancing of systems, and system turnover.

Compact fluorescent lamp (CFL): A fluorescent bulb that is designed to fit in a normal light fixture. They use less energy and last longer than incandescent bulbs.

Conservation: Using natural resources wisely and at a slower rate than normal.

Corporate Sustainability Report (CSR): An annual report by a company documenting its efforts to reduce negative environmental and social impacts.

Deconstruction: Taking a building apart with the intent of salvaging reusable materials.

Deforestation: The removal of trees without sufficient replanting.

Desertification: The creation of deserts through degradation of productive land in dry climates by human activities.

Downcycling: Recycling one material into a material of lesser quality. An example is the recycling of plastics into lower-grade composites.

EarthCraft: A residential green building program of the Greater Atlanta Home Builders Association in partnership with Southface Energy Institute.

Ecological footprint: A measure of impact human activities have on the environment. It compares human consumption of natural resources with the Earth's capacity to regenerate them. Measured in hectares or acres.

Ecosystem: A combination of all plants, animals and microorganisms in an area that complement each other. These function together with all of the nonliving physical factors of the environment.

Embodied energy: The total energy required to bring a product to market. It includes raw material extraction, manufacturing, final transport, and installation.

Endangered species: A population of a species at risk of becoming extinct. A threatened species is any species that is vulnerable to extinction in the near future.

Energy efficiency: Getting more use out of electricity already generated.

ENERGY STAR: A United States government program to promote energy efficiency. It is a joint program of the US Environmental Protection Agency and the US Department of Energy.

Environmental Product Declaration (EPD): A document listing the environmental impacts caused by manufacture of a product.

Equipment idling: The operation of equipment while it is not in motion or performing work. Limiting idle times reduces air pollution and greenhouse gas emissions.

Erosion: The displacement of solids by wind, water, ice, or gravity or by living organisms. These solids include rocks and soil particles.

Flood plain: An area surrounding a river or body of water that regularly floods within a given period of time.

Flushout: Using fresh air in the building HVAC system to remove contaminants from the building.

FSC Certified: Wood or wood products that have met the Forest Stewardship Council's tracking process for sustainable harvest.

Fugitive emissions: Pollutants released to the air other than those from stacks or vents. They are often due to equipment leaks, evaporative processes, and wind disturbances.

Geothermal: Heat that comes from within the Earth.

Global climate change: Changes in weather patterns and temperatures on a planetary scale. This may lead to a rise in sea levels, melting of polar ice caps, increased droughts, and other weather effects.

Graywater: A nonindustrial wastewater generated from domestic processes. These include laundry and bathing. Graywater comprises 50 to 80 percent of residential wastewater.

Greenhouse Effect: The overall warming of a planet's surface due to retention of solar heat by its atmosphere.

Green Seal Certified: A certification of a product to indicate its environmental friendliness. Green Seal is a group that works with manufacturers, industry sectors, purchasing groups, and government at all levels to green the production and purchasing chain. Founded in 1989, Green Seal provides science-based environmental certification standards.

Grid intertie: A connection between a local source of power and the utility power grid.

Halons: Ozone-depleting compounds consisting of bromine, fluorine, and carbon. They were commonly used as fire extinguishing agents, both in built-in systems and in handheld portable fire extinguishers. Halon production in the US ended on December 31, 1993.

Hardscape: Paved areas surrounding a project, including parking lots and sidewalks.

Heat island: An area that is warmer than surrounding areas due to absorbing more solar radiation.

Hybrid vehicle: A vehicle that uses two or more power sources for propulsion.

Hydrochlorofluorocarbons (HCFCs): A group of man-made compounds containing hydrogen, chlorine, fluorine, and carbon. They are used for refrigeration, aerosol propellants, foam manufacture, and air conditioning. They are broken down in the lowest part of the atmosphere and pose a smaller risk to the ozone layer than other types of refrigerants.

Hydrologic cycle: The circulation and conservation of Earth's water supply. The process has five phases: condensation, infiltration, runoff, evaporation, and precipitation.

Indoor air quality (IAQ): The content of interior air that could affect health and comfort of building occupants.

Infiltration: The movement of air from outside a building to inside through cracks or openings in the building envelope.

Insulating concrete form (ICF): Rigid forms that hold concrete in place during curing and remain in place afterwards. The forms serve as thermal insulation for concrete walls.

Integrative design: A collaborative design methodology that emphasizes the input of knowledge from several areas in the development of a complete design.

Just-in-time delivery: A material delivery strategy that reduces material inventory. The material that is delivered is used immediately.

Leadership in Energy and Environmental Design (LEED):: A point-based rating system used to evaluate the environmental performance of buildings. Developed by the US Green Building Council, LEED provides a suite of standards for environmentally sustainable construction. Founded in 1998, LEED focuses on the certification of commercial and residential buildings and the neighborhoods in which they exist.

Life cycle: The useful life of a system, product, or building.

Life cycle assessment: An analytic technique to evaluate the environmental impact of a system, product, or building throughout its life cycle. This includes the extraction or harvesting of raw materials through processing, manufacture, installation, use, and ultimate disposal or recycling.

Life cycle cost: The cost of a system or a component over its entire life span.

Light-emitting diode (LED) lamp: A highly efficient, electronic light bulb using a glowing diode designed to fit a standard light fixture.

Lithium ion (Li-ion): A type of rechargeable battery in which a lithium ion moves between the anode and cathode. They are commonly used in consumer electronics.

Local materials: Materials that come from within a certain number of miles from the project. Materials produced locally use less energy during transportation to the site. The LEED system of building certification offers points for the use of regional or local materials.

Location valuation: A bonus under the LEED rating system for using a product produced locally. Materials from within a 100-mile radius can be counted as twice their value as a reward for using local products.

Low-emission vehicle:: Vehicles that produce fewer emissions than the average vehicle. Beginning in 2001, all light vehicles sold nationally were required to meet this standard.

Maintainability: An indication of how difficult it is to properly maintain a building system or technology.

Massing: The three-dimensional shape of a building, including length, width, and height. Determines the amount of a building exposed to sunlight.

Minimum efficiency reporting value (MERV): A measurement scale designed in 1987 by the American Society of Heating, Refrigeration, and Air-Conditioning Engineers (ASHRAE) to rate the effectiveness of air filters. The scale is designed to represent the worst-case performance of a filter when dealing with particles in the range of 0.3 to 10 microns.

Monofill: A landfill that accepts only one type of waste, typically in bales.

Multi-function material: A material that can be used to perform more than one function in a facility.

Nano-material: A material with features smaller than a micron in at least one dimension.

Nickel cadmium (NiCad): A popular type of rechargeable battery using nickel oxide hydroxide and metallic cadmium as electrodes.

Nonrenewable: A material or energy source that cannot be replenished within a reasonable period.

Nontoxic: Substances that are not poisonous.

Offgassing: The evaporation of volatile chemicals at normal atmospheric pressure. Building materials release chemicals into the air through evaporation.

Ozone depletion: A slow, steady decline in the total amount of ozone in Earth's stratosphere.

Ozone hole: A large, seasonal decrease in stratospheric ozone over Earth's polar regions. The ozone hole does not go all the way through the layer.

Papercrete: A fiber-cement material that uses waste paper for fiber.

Passive solar design: Designing a building to use the sun's energy for lighting, heating, and cooling with minimal additional inputs of energy.

Passive survivability: The ability of a building to continue to offer basic function and habitability after a loss of infrastructure (i.e., water and power).

Pathogen: An infectious agent that causes disease or illness.

Payback period: The amount of time it takes to break even on an investment; the period of time after which savings from an investment equals its initial cost.

Peak shaving: An energy management strategy that reduces demand during peak times of the day and shifts it to off-peak times, such as at night.

Persistent, bioaccumulative toxin (PBT): A harmful substance such as a pesticide or organic chemical that does not quickly degrade in the natural environment but does accumulate in plants and animals in the food chain.

Pervious concrete: A mixture of coarse aggregate, Portland cement, water, and little to no sand. It has a 15 to 25 percent void structure and allows 3 to 8 gallons of water per minute to pass through each square foot. Also known as permeable concrete.

Phantom loads: Electricity consumed by appliances and devices when they are switched off.

Phase change material (PCM): A material that stores or releases large amounts of energy when it changes between solid, liquid, or gas forms.

Photosensor: A sensor that measures daylight and adjusts artificial lighting to save energy.

Photovoltaic: A technology that converts light directly into electricity.

Pollution prevention: The prevention or reduction of pollution at the source.

Post-consumer: Products made out of material that has been used by the end consumer and then collected for recycling.

Post-industrial/pre-consumer: Material diverted from the waste stream during the manufacturing process.

Preservation: The act and advocacy of protecting the natural environment.

Radio-frequency identification (RFID): An automatic identification method. It relies on storing and remotely retrieving data using devices called RFID tags.

Rainwater harvesting: The gathering and storing of rainwater.

Rammed earth: An ancient building technique similar to adobe using soil that is mostly clay and sand. The difference is that the material is compressed or tamped into place, usually with forms that create very flat vertical surfaces.

Rapidly renewable: A material that is replenished by natural processes at a rate comparable to its rate of consumption. The LEED system of building certification rewards use of rapidly renewable materials that regenerate in 10 years or less, such as bamboo, cork, wool, and straw.

Raw material: A material that has been extracted directly from nature. It is in an unprocessed or minimally processed state.

Recyclable: Material that still has useful physical or chemical properties after serving its original purpose. It can be reused or remanufactured into additional products. Plastic, paper, glass, used oil, and aluminum cans are examples of recyclable materials.

Recycled content: A material containing components that would otherwise have been discarded.

Recycled plastic lumber (RPL): A wood-like product made from recovered plastic, either by itself or mixed with other materials. It can be used as a substitute for concrete, wood, and metals.

Recycling: The reprocessing of old materials into new products. A goal is to prevent the waste of potentially useful materials and reduce the consumption of new materials.

Renewable: A resource that may be naturally replenished.

Reusable: A material that can be used again without reprocessing. This can be for its original purpose or for a new purpose.

Salvaged: A used material that is saved from destruction or waste by being reused.

Sedimentation: The process of depositing a solid material from a state of suspension in a fluid, usually air or water.

Sick Building Syndrome: A variety of illnesses thought to be caused by poor indoor air. Symptoms include headaches, fatigue, and other problems that increase with continued exposure.

Smart material: Materials that have one or more properties that can be significantly changed. These changes are driven by external stimuli.

Softscape: The area surrounding a building that contains plantings, lawn, and other vegetated areas.

Solar: Energy from the sun in the form of heat and light.

Solid waste: Products and materials discarded after use in homes, businesses, restaurants, schools, industrial plants, or elsewhere.

Solvent-based: A material that consists of particles suspended or dissolved in a solvent. A solvent is any substance that will dissolve another. Solvent-based building materials typically use chemicals other than water as their solvent, including toluene and turpentine, with hazardous health effects.

Spoil pile: Excavated soil that has been moved and temporarily stored during construction.

Sprawl: Unplanned and inefficient development of open land.

Stormwater runoff: Unfiltered water that reaches streams, lakes, and oceans after a rainstorm by flowing across impervious surfaces.

Strawbale construction: A building method that uses straw as the structural element, insulation, or both. It has advantages over some conventional building systems because of its cost and availability.

Structural insulated panel (SIP): A composite building material used for exterior building envelopes. It consists of a sandwich of two layers of structural board with an insulating layer of foam in between. The board is usually oriented strand board (OSB) and the foam can be polystyrene, soy-based foam, urethane, or even compressed straw.

Sustainably harvested:: A method of harvesting a material from a natural ecosystem without damaging the ability of the ecosystem to continue to produce the material indefinitely.

Takeback:: A condition where manufacturers recover waste from packaging or products after use, at the end of their life cycle.

Thermal bridging:: A condition created when a thermally conductive material bypasses an insulation system, allowing the rapid flow of heat from one side of a building wall to the other. Metal components, including metal studs, nails, and window frames, are common culprits.

Thermal mass:: A property of a material related to density that allows it to absorb heat from a heat source, and then release it slowly. Common materials used to provide thermal mass include adobe, mud, stones, or even tanks of water.

Urbanization:: Converting rural land to higher density development.

Urea formaldehyde:: A transparent thermosetting resin or plastic. It is made from urea and formaldehyde heated in the presence of a mild base. Urea formaldehyde has negative effects on human health when allowed to offgas or burn.

Vapor-resistant: A material that resists the flow of water vapor.

Virgin material: A material that has not been previously used or consumed. It also has not been subjected to processing. See also *raw material*.

Volatile organic compounds (VOCs): Gases that are emitted over time from certain solids or liquids. Concentrations of many VOCs are up to 10 times higher indoors than outdoors. Examples include paints and lacquers, paint strippers, cleaning supplies, pesticides, building materials, and furnishings.

Walk-off mat: Mats in entry areas that capture dirt and other particles.

Waste separation: Sorting waste by specific type of material and storing it in different containers to facilitate recycling.

Water efficiency: Managing water use to prevent waste or overuse. Includes using less water to achieve the same benefits.

Water footprint: The volume of water used directly or indirectly to sustain something over a period of time.

Water-based: Materials that uses water as a solvent or vehicle of application.

Water-resistant: A material that hinders the penetration of water.

Waterproof: A material that is impervious to or unaffected by water.

Wetland: Lands where saturation with water is the dominant factor. This determines the way soil develops and the types of plant and animal communities living in the soil and on its surface.

Xeriscaping: Landscaping that requires little or no water for irrigation. Xeriscaping can be achieved through smart plant selection, mulching, and other tactics.

Zoning: Rules for placing similar items next to one another, as in plants within a landscape, or types of buildings within a community.

COMMON ACRONYMS

AAC	Aerated autoclaved concrete
ACH	Air changes per hour
BIPV	Building-integrated photovoltaic
BMP	Best management practice
BTU	British thermal unit
CFC	Chlorofluorocarbon
CFL	Compact fluorescent lamp
CO	Carbon monoxide
CO_2	Carbon dioxide
CSR	Corporate Sustainability Report
EA	Energy and Atmosphere (LEED credit category)
EEM	Energy efficient mortgage
EMF	Electromagnetic field
EPA	Environmental Protection Agency
EPD	Environmental Product Declaration
EQ	Indoor Environmental Quality (LEED credit category)
ERV	Energy recovery ventilator
FSC	Forest Stewardship Council
GBCI	Green Building Certification Institute
GHG	Greenhouse gas
HCFC	Hydrochlorofluorocarbon
HERS	Home energy rating system
HID	High intensity discharge
HVFA	High volume flyash
IAQ	Indoor air quality
ICF	Insulating concrete form
ID	Innovation in Design (LEED credit category)
IDP	Integrative design process
ISO	International Standards Organization
LCA	Life cycle assessment
LEED	Leadership in Energy and Environmental Design
LEED-AP	Leadership in Energy and Environmental Design Accredited Professional
LEED-BD+C	Leadership in Energy and Environmental Design for Building Design and Construction and Major Renovation
LEED-ID+C	Leadership in Energy and Environmental Design for Interior Design and Construction
LEED-O+M	Leadership in Energy and Environmental Design for Operations and Maintenance of Existing Buildings
LEED-H	Leadership in Energy and Environmental Design for Homes
LEED-ND	Leadership in Energy and Environmental Design for Neighborhood Development

LED	Light-emitting diode
Li-ion	Lithium ion
Low-E	Low-emissivity
LT	Location and Transportation (LEED credit category)
MCS	Multiple chemical sensitivity
MDF	Medium-density fiberboard
MERV	Minimum efficiency reporting value
MPR	Minimum Program Requirements
MR	Materials and Resources (LEED credit category)
N_2O	Nitrous oxide
NACH	Natural air changes per hour
NAHB	National Association of Home Builders
NFRC	National Fenestration Rating Council
NiCad	Nickel cadmium
OPC	Off-peak cooling
OSB	Oriented strand board
OVE	Optimum value engineering
PB	Particleboard
PBT	Persistent bioaccumulative toxin
PCM	Phase change material
PF	Phenyl formaldehyde
PtD	Prevention through design
PV	Photovoltaic
PVC	Polyvinyl chloride
REC	Renewable Energy Credit
RFID	Radio-frequency identification
RP	Regional Priority (LEED credit category)
RPL	Recycled plastic lumber
R-value	Resistance to heat flow
SEER	Seasonal Energy Efficiency Ratio
SFI	Sustainable Forest Initiative
SHGC	Solar heat gain coefficient
SIP	Structural insulated panel
SMACNA	Sheet Metal and Air Conditioning Contractors' National Association
SS	Sustainable Sites (LEED credit category)
TPO	Thermoplastic polyolefin
UF	Urea formaldehyde
USGBC	United States Green Building Council
UV	Ultraviolet
U-value	Resistance to heat loss (also known as U-factor)
VOC	Volatile organic compound
WE	Water Efficiency (LEED credit category)

WHAT IT MEANS TO BE GREEN

What makes a building green? Green buildings are designed to use resources efficiently. This includes energy, water, and building materials. Green buildings must also provide a healthy, comfortable, and safe environment for the occupants. They should be constructed in a way that minimizes their impact on the environment. The qualities that make a building green depend on the systems, methods, equipment, and materials used in its construction.

Craft professionals make a large impact on the green environment simply by how they use materials. Do you know that the building industry generates over thirty million tons of construction waste each year that ends up in landfills? Yet almost eighty percent of that waste could be recycled. Think about the waste that you generate. Can that copper wire be recycled? What about the cast iron or PVC? Is the waste on your project recycled? What happens to excess materials at the end of the project? Are they recovered, or are they thrown away?

Carpenters, electricians, plumbers, and HVAC technicians all play significant roles in the greening of the built environment. As a craft worker, you must be aware of the things you do every day that help to make a building green, and be willing to learn new methods and work with new technologies.

This section provides an overview of what it means to be a green craft professional. Whatever your trade, there are specific methods that can be applied towards building more energy-efficient commercial and residential structures with less environmental impact.

WHAT DOES IT MEAN TO BE A GREEN CARPENTER?

As a carpenter, you contribute to the efficiency of a building in many ways. For example, the method of framing you use can improve the energy efficiency of a building. The materials you install on the exterior can improve the durability of the building. You can conserve resources by reducing the amount of material used to build a structure. You can also recycle waste materials. Your daily activities on a job site affect the efficiency of a building.

The emphasis on efficiency has led to changes in materials and construction methods. One of these methods is Optimum Value Engineering (OVE), a resource-efficient method of framing. This method allows for greater energy efficiency and reduction in construction costs. It uses a single top plate, eliminates headers in interior walls, and uses wider stud spacing. This method consumes less wood and creates more space for insulation in exterior walls. All of this directly improves the efficiency of the structure. Learning and using the principles of OVE is a direct way that carpenters help the green environment.

The green movement has also been marked by the use of building components for various purposes. A roofing system is an example. The roof may be designed to do more than protect the occupants from the elements. It may be connected to a rainwater collection system, or it may be a green or vegetated roof. These non-traditional systems require the skills of carpenters to construct major components.

Over the last few decades, carpenters have started to use new greener building materials. An example is the supplementing of traditional dimensional lumber used in framing with engineered lumber and I-joists. More recently, multi-function building materials have been introduced. These include structural insulated panels and insulated concrete forms, each providing structural support and insulation. These materials will change the way carpenters frame structures and build foundations. Other green building materials include insulated foam sheathing and recycled flooring. These innovations require carpenters to successfully use them as part of the green strategy.

You should also be aware of the types of materials being used on a job site. If a project is seeking to be green, be sure to look for the Forest Stewardship Council (FSC) certification label on lumber. If the lumber on the job site does not have an FSC seal, alert your supervisor. Another way you can be green-minded is to read labels. Check the recycled content of materials you are using. Also check the labels on adhesives to determine if they are low VOC or non-toxic. These steps promote green building.

What does it mean to be a Green Electrician?

As an electrician, you will contribute significantly to the success of an energy-efficient building. Both residential and commercial buildings consume large amounts of energy. This energy is used in lighting, heating, cooling, and operating the building. Building owners are looking for ways to reduce their energy bills through conservation and efficiency. The equipment you install can directly assist with this. It can also produce energy more responsibly. You can help by carefully following specifications when installing fixtures, and installing only the minimum specified. This reduces energy use and can directly reduce the amount of greenhouse gases generated from coal-fired power plants. Your daily activities on a job site affect the efficiency of a building. They also impact the environment.

The emphasis on energy efficiency has led to innovations in energy conservation in buildings. This includes energy management systems. These systems have advanced controls that automatically adjust lighting during the day. They also include environmental controls that manage the temperature and humidity of a building. Electronic sensors and meters are also used to monitor energy use and building conditions. These systems work together to ensure the comfort of the occupants. Some buildings have controls that can be used directly by occupants. Learning how to install and trouble-shoot these systems is a direct way that electricians help the green environment.

The green movement also supports the use of alternative energy systems. Alternative energy systems include sources such as solar, wind, and biomass. Solar photovoltaic panels are an example of an alternative energy system. The installation of these panels requires skilled electricians who know how to wire them correctly. Alternative energy sources need proper power distribution to work well. These non-traditional systems require the skills of electricians to correctly install and maintain major components.

Electricians have adapted to new greener building technologies and methods. An example is the reduction of incandescent lighting fixtures. Electricians have been installing fluorescent lighting and replacing the ballast in these fixtures for decades. More recently, solid state lighting including light-emitting diodes is becoming popular. This is due to their energy efficiency and durability.

Light pollution reduction is also a critical area for green building. The correct installation and location of outdoor lighting fixtures can help limit light pollution. Each of these innovations requires electricians to successfully use them as part of the green strategy. Remember, careful and proper disposal of ballasts and fluorescent lamps is important, as these contain hazardous materials.

Methods are just as important. Using only the amount of wire needed in a building is a way to use resources efficiently. Another example is to be sure not to disturb the insulation in the wall when replacing wiring or installing new electrical boxes. This will cause gaps in the insulation and reduce the efficiency of the building. Taking these steps on the job site will help to maintain the energy efficiency of a building.

WHAT DOES IT MEAN TO BE A GREEN HVAC TECHNICIAN?

As an HVAC technician, your work has a large impact on the comfort and health of building occupants. Your work also impacts the ability of a building to be green. Heating, cooling, and ventilation consume the largest amount of energy in a building. This energy consumption increases the burning of fossil fuel and greenhouse gas emissions. HVAC systems designed, installed, commissioned, and maintained properly will reduce these impacts. This is where HVAC technicians can make the greatest impact on the green environment.

A challenge facing green HVAC technicians is the correct sizing of air ducts. Another challenge is determining the conditioning capacity required in a facility. The rule of thumb based on square footage is not good enough for modern green facilities. Correct sizing is critical for energy efficiency and the comfort of the occupants. You will begin to see a shift in the size and types of equipment and ductwork you install.

Over the past decade, there has been a shift in the types of equipment used for heating and air conditioning. These include high efficiency heat pumps and air conditioning units with higher Seasonal Energy Efficiency Ratios (SEER). If a project is seeking to be green, then be sure to look for the appropriate ratings on equipment. If the air conditioning unit does not have the appropriate SEER rating, alert your supervisor. Another way to be green-minded is to be sure you are installing enough returns to keep a facility balanced. If you know there are not enough returns, tell your supervisor. Keep in mind the new greener building codes. In today's building environment, HVAC technicians are responsible for ensuring the appropriate equipment is installed.

Conserving resources and efficiently using materials go hand-in-hand for HVAC systems. How and where ductwork is installed can save energy and materials. An example is compact air distribution systems. These tend to keep ductwork runs as short as possible to minimize the amount of ductwork needed. Another example is to lay out ductwork so air can be discharged from inside walls or ceilings. This may also reduce the amount of ductwork that runs through unconditioned space. If it must be run through unconditioned space for practical purposes, be sure to insulate the ducts, as this improves energy efficiency.

Quality workmanship is critical to maintain healthy indoor air quality. No other trade has a more direct impact on air quality than HVAC. Sealing ductwork is extremely important. Improperly sealed return air ducts draw in humidity that may promote mold growth. Also, leaking ducts cause air handlers to work harder and use more energy. Unsealed return ducts may also draw dust or harmful gases such as carbon monoxide into conditioned space. This may place the health of the occupants at risk. Unconnected ductwork can cause pressure differences in a building. This difference may cause conditioned air to move through the building envelope. This will decrease the efficiency of any facility and increase energy use. As an HVAC technician in the green environment, you will help ensure occupant health and safety and the energy efficiency of a building.

HVAC technicians also play a critical role in protecting the green environment. Refrigerants used in air conditioning equipment can damage the ozone layer if allowed to leak. Antifreeze and other chemicals leaked to the soil can be harmful to the water supply. As an HVAC technician in the green environment, you will help maintain the health of the surrounding community and world at large.

In addition to improved construction methods, new technologies are being used to help improve the efficiency of buildings. An example is Energy Recovery Ventilation (ERV). During the heating season, this system recovers heat from indoor air exhausts and then transfers this heat to fresh incoming air. In the cooling season, it transfers heat from incoming fresh air to the exhaust. This reduces the amount of energy required to condition fresh air. These new systems will be part of greener buildings, and installation will require your experience.

Energy-efficient HVAC systems will require involvement of contractors early in the design phase of a building. This will help to ensure that owners understand how important the location of equipment and ductwork is for energy efficiency. They will also need to know the benefits of high efficiency heat pumps and air conditioners. During the building's life, they will need the expertise of HVAC technicians to maintain and recommission systems to ensure efficient operation. As a technician, you will install the equipment, ductwork, and returns indicated in the design. The days of on-site location of ducts is soon to be a thing of the past in the green workforce.

What does it mean to be a Green Plumber?

As a plumber, your work will affect the efficient use of water in a building. By installing and maintaining water-efficient systems, plumbers help reduce energy costs. The EPA estimates that three to four percent of all energy consumed in the United States is associated with treating drinking water and wastewater. Water-efficient systems directly help reduce energy use and greenhouse gas emissions. Gas emissions in homes and businesses can occur if water heaters and other combustion fixtures are not installed properly. Your work will directly impact the health and safety of the occupants.

The emphasis on efficiency has led to new innovations in water and energy conservation. The use of tankless hot water heaters is an example of a way to save energy in a building. Solar hot water heating and drain water heat recovery systems are examples, too. Many building owners are installing low-flow or dual-flush toilets and waterless urinals to reduce water use. These fixtures and systems require a plumber who is knowledgeable about their installation and maintenance.

Green technologies are shaping the future of water conservation efforts. Graywater systems are becoming more popular as new building codes allow their use. Plumbers will be responsible for installing the piping network for these systems. Rainwater harvesting systems are another example. Proper sizing of pipes, storage tanks, and collection systems may be your responsibility. On-site wastewater treatment systems are becoming more common as an alternative to traditional sewage connections. These systems require special piping and on-site treatment vessels. The effective design of plumbing systems for green roofs will also be an important role for green plumbers. These non-traditional plumbing systems require the skills of plumbers to correctly install and maintain major components.

Plumbers have adapted to new greener building technologies. An example is the increase in low-flow fixtures being installed. Sensor-operated fixtures save water and reduce the transmission of germs between users. These technologies require plumbers to successfully install and maintain them as part of the green strategy. They also require careful system design to be sure water isn't kept too long in pipes. This increases water age and can cause water quality problems for building users.

The greener methods are important too. Use your resources wisely by running the shortest length of pipe required. Consider the size of the cuts you make in the floors and walls when installing fixtures. Your cuts can create gaps that allow heating and cooling energy to be wasted. Be sure to make holes the appropriate size and always seal them. Taking these steps on a job site will help to maintain the energy efficiency of a building.

LEED FOR BUILDING DESIGN + CONSTRUCTION, VERSION 4.0
REGISTERED PROJECT CHECKLIST

Location and Transportation		16
Credit	LEED for Neighborhood Development Location	16
Credit	Sensitive Land Protection	1
Credit	High Priority Site	2
Credit	Surrounding Density and Diverse Uses	5
Credit	Access to Quality Transit	5
Credit	Bicycle Facilities	1
Credit	Reduced Parking Footprint	1
Credit	Green Vehicles	1

Sustainable Sites		10
Prereq	Construction Activity Pollution Prevention	Required
Credit	Site Assessment	1
Credit	Site Development - Protect or Restore Habitat	2
Credit	Open Space	1
Credit	Rainwater Management	3
Credit	Heat Island Reduction	2
Credit	Light Pollution Reduction	1

Water Efficiency		11
Prereq	Outdoor Water Use Reduction	Required
Prereq	Indoor Water Use Reduction	Required
Prereq	Building-Level Water Metering	Required
Credit	Outdoor Water Use Reduction	2
Credit	Indoor Water Use Reduction	6
Credit	Cooling Tower Water Use	2
Credit	Water Metering	1

Energy and Atmosphere		33
Prereq	Fundamental Commissioning and Verification	Required
Prereq	Minimum Energy Performance	Required
Prereq	Building-Level Energy Metering	Required
Prereq	Fundamental Refrigerant Management	Required
Credit	Enhanced Commissioning	6
Credit	Optimize Energy Performance	18
Credit	Advanced Energy Metering	1
Credit	Demand Response	2
Credit	Renewable Energy Production	3
Credit	Enhanced Refrigerant Management	1
Credit	Green Power and Carbon Offsets	2

Materials and Resources		**13**
Prereq	Storage and Collection of Recyclables	Required
Prereq	Construction and Demolition Waste Management Planning	Required
Credit	Building Life-Cycle Impact Reduction	5
Credit	Building Product Disclosure and Optimization - Environmental Product Declarations	2
Credit	Building Product Disclosure and Optimization - Sourcing of Raw Materials	2
Credit	Building Product Disclosure and Optimization - Material Ingredients	2
Credit	Construction and Demolition Waste Management	2

Indoor Environmental Quality		**16**
Prereq	Minimum Indoor Air Quality Performance	Required
Prereq	Environmental Tobacco Smoke Control	Required
Credit	Enhanced Indoor Air Quality Strategies	2
Credit	Low-Emitting Materials	3
Credit	Construction Indoor Air Quality Management Plan	1
Credit	Indoor Air Quality Assessment	2
Credit	Thermal Comfort	1
Credit	Interior Lighting	2
Credit	Daylight	3
Credit	Quality Views	1
Credit	Acoustic Performance	1

Innovation		**6**
Credit	Innovation	5
Credit	LEED Accredited Professional	1

Regional Priority		**4**
Credit	Regional Priority: Specific Credit	1
Credit	Regional Priority: Specific Credit	1
Credit	Regional Priority: Specific Credit	1
Credit	Regional Priority: Specific Credit	1

TOTAL Possible Points:		**110**

Version published on 6 June 2014

WORKSHEET 1: INVENTORY YOUR HOUSEHOLD IMPACTS

WORKSHEET 1: INVENTORY YOUR HOUSEHOLD IMPACTS

Conducting an inventory of your household consumption, waste generation, and activities is the first step in understanding how you can reduce your impact. Answer the following questions based on your best guess. If you share a household with other people, divide the total answer for your household by the number of people who live in your home.

How many gallons of garbage do you throw away each week? Find out the size of your trash bin in liters. Multiply by 52 to calculate the liters of garbage you throw away per year.

Liters of garbage per year: _____

How much electricity do you use per year? You'll find this information on your electricity bill. You can add up the total for 12 months worth of bills, or multiply a monthly average by 12. Most energy bills tell you usage in kilowatt hours. If your bill uses a different unit, you can look up a unit converter at www.converterunits.com.

Total electricity per year in kilowatt-hours: _____

How much natural gas do you use per year? Check your natural gas bill if you get one. It will tell you how many therms of gas you use per month. Your highest values will probably be during the winter heating season. If you know your monthly average, multiply by 12 to calculate your annual use. In the United States, the unit for natural gas is the therm. If your bill uses a different unit, you can look up a unit converter at www.converterunits.com.

Total therms of natural gas per year: _____

How many liters of propane do you use each year? You can check this on your propane bill if you get one. Add up the total number of liters per year.

Total liters of propane per year: _____

How many liters of fuel oil do you use per year? You can check this on your fuel oil bill if you get one. Add up the total number of liters per year.

Total liters of fuel oil per year: _____

On average, what is your monthly combined water and sewage bill? Check your monthly bill if you get one and add up all the amounts for a one-year period. If you don't get a water and sewage bill, you can estimate this amount as approximately $75 USD per person per year. You can look up a currency converter to convert your currency to US dollars at www.xe.com.

Annual cost in US dollars for water and sewage: _____

How many square meters is your house or dwelling? If you don't know, draw a floor plan of your house or apartment and estimate the floor area in square meters.

Size of house (floor area) in square meters: _____

70101-14_A01A.EPS

WORKSHEET 1 (Continued)

On average, how many kilometers do you drive your household vehicles per week, and what are their average fuel efficiencies? 483 kilometers per week per vehicle or 24,140 kilometers per year is about average in the United States. If you're not sure about fuel efficiency, assume your car gets 9.4 kilometers per liter, which is about average. If you know how many liters of fuel you use per week, skip directly to the end of the line.

Car 1 kilometers per week _____ / Car 1 kilometers per liter _____ = Car 1 liters per week _____

Car 2 kilometers per week _____ / Car 2 kilometers per liter _____ = Car 2 liters per week _____

Car 3 kilometers per week _____ / Car 3 kilometers per liter _____ = Car 3 liters per week _____

Add up the liters of gas per week for all your vehicles, and then multiply by 52 to estimate the liters of gasoline you use per year.

Total liters of gasoline you use per year: _____

On average, how much do you travel each year on airplanes? Estimate the number of flight segments below for each of the three distances. A round trip counts as two segments, and each segment of a multi-segment flight counts as its own flight. For example, if you fly from Baltimore to Cincinnati to Atlanta, that counts as two flight segments.

Number of short-haul flight segments (less than 1,126 kilometers or 2 hours): _____

Number of medium-haul flight segments (1,126 – 4,023 kilometers or 2 – 4 hours): _____

Number of long-haul flight segments (more than 4,023 kilometers or longer than 4 hours): _____

On average, how many kilometers do you travel on public transportation per year?

Number of kilometers per year on transit bus/subway: _____

Number of kilometers per year on intercity bus: _____

Number of kilometers per year on intercity train: _____

Add up the total number of *kilometers* per year on public transportation: _____

Your answers to these questions will help you calculate your own carbon footprint later in the module. Keep track of your answers in the workbook or on a separate worksheet.

70101-14_A01B.EPS

WORKSHEET 2: INVENTORY YOUR PRODUCT IMPACTS

WORKSHEET 2: INVENTORY YOUR PRODUCT IMPACTS

Consider the products you buy or that are bought for you over the course of a year. Answer the following questions based on your best guess. If you share a household with other people, divide the total answer for your household by the number of people who live in your home. You can use your own currency to estimate how much you spend on each item. You will need to convert to US dollars before you calculate your carbon footprint. Find your currency's US dollar equivalent at www.xe.com.

1 of your currency = $ _____ US dollars. Write this amount in the third space in each line below to convert your currency to US dollars.

Eating out: $ _____ *per month* × 12 = _____ *per year* × _____ $US currency = $ _____ *per year*

Meat, fish, & protein: $ _____ *per month* × 12 = _____ *per year* × _____ $US currency = $ _____ *per year*

Cereals & baked goods: $ _____ *per month* × 12 = _____ *per year* × _____ $US currency = $ _____ *per year*

Dairy: $ _____ *per month* × 12 = _____ *per year* × _____ $US currency = $ _____ *per year*

Fruits & vegetables: $ _____ *per month* × 12 = _____ *per year* × _____ $US currency = $ _____ *per year*

Other: $ _____ *per month* × 12 = _____ *per year* × _____ $US currency = $ _____ *per year*

On average each month, how much do you spend on the following goods and services? Multiply each amount by 12 to estimate your average annual spending in each category. If you already know how much you spend per year, you can skip the monthly amount and write the average amount per year at the end of each line. Don't forget to convert to US dollars at the end.

Clothing: $ _____ *per month* × 12 = _____ *per year* × _____ $US currency = $ _____ *per year*

Furnishings &
   household items: $ _____ *per month* × 12 = _____ *per year* × _____ $US currency = $ _____ *per year*

Other goods: $ _____ *per month* × 12 = _____ *per year* × _____ $US currency = $ _____ *per year*

Services: $ _____ *per month* × 12 = _____ *per year* × _____ $US currency = $ _____ *per year*

70101-14_A02.EPS

WORKSHEET 3: DETERMINE YOUR CARBON FOOTPRINT

WORKSHEET 3: DETERMINE YOUR CARBON FOOTPRINT

Fill in the amounts from Worksheets 1 and 2 to calculate your total carbon footprint in kilograms of carbon. Which of your activities contributes the most to your carbon footprint?

Item	Quantity	Carbon Factor	Total Kilograms of CO_2/Year
Liters of garbage/year:	_____	× 0.24 kg per liter	= _____ kg per year
Total electricity/year in kilowatt-hours:	_____	× 0.64 kg per kWh	= _____ kg per year
Total therms of natural gas/year:	_____	× 1.40 kg per liter	= _____ kg per year
Total liters of propane/year:	_____	× 1.52 kg per liter	= _____ kg per year
Total liters of fuel oil/year:	_____	× 2.68 kg per liter	= _____ kg per year
Annual cost (USD) for water and sewage:	_____	× 4.04 kg per dollar	= _____ kg per year
House size (floor area) in square meters:	_____	× 10.25 sq m	= _____ kg per year
Liters of gasoline/year:	_____	× 2.40 kg per liter	= _____ kg per year
No. of short-haul flight segments/year:	_____	× 137.89 kg per segment	= _____ kg per year
No. of medium-haul flight segments/year:	_____	× 329.31 kg per segment	= _____ kg per year
No. of long-haul flight segments/year:	_____	× 1,005.61 kg per segment	= _____ kg per year
Kilometers per year on public transportation:	_____	× 0.14 kg per km	= _____ kg per year
Eating out (US dollars/year):	_____	× 0.36 kg per dollar	= _____ kg per year
Meat, fish, & protein (US dollars/year):	_____	× 1.45 kg per dollar	= _____ kg per year
Cereals & baked goods (US dollars/year):	_____	× 0.73 kg per dollar	= _____ kg per year
Dairy (US dollars/year):	_____	× 1.91 kg per dollar	= _____ kg per year
Fruits & vegetables (US dollars/year):	_____	× 1.18 kg per dollar	= _____ kg per year
Other (US dollars/year):	_____	× 0.45 kg per dollar	= _____ kg per year
Clothing (US dollars/year):	_____	× 0.45 kg per dollar	= _____ kg per year
Household items (US dollars/year):	_____	× 0.45 kg per dollar	= _____ kg per year
Other goods (US dollars/year):	_____	× 0.34 kg per dollar	= _____ kg per year
Services (US dollars/year):	_____	× 0.18 kg per dollar	= _____ kg per year

Total Carbon Footprint: = _____ kg per year

70101-14_A03.EPS

Additional Resources

This module presents thorough resources for task training. The following resource material is suggested for further study.

Field Guide for Sustainable Construction. Department of Defense – Pentagon Renovation and Construction Office. (2004). PDF, 2.6 MB, 312 pgs. Available for download at **www.wbdg.org**.

Greening Federal Facilities, 2nd Ed. US Department of Energy Federal Energy Management Program. (2001). PDF, 2.1 MB, 211 pgs. Available for download at **www.wbdg.org**.

Natural Capitalism. Lovins, A., Hawkin, P., and Lovins, L.H. (1995). Little, Brown, & Company, Boston, MA. Available online at **www.natcap.org**.

Sustainable Buildings Technical Manual. Public Technologies, Inc./U.S. Department of Energy. (2006). PDF, 3.1 MB, 292 pgs. Available for download at **www.greenbiz.com**.

Sustainable Buildings and Infrastructure: Paths to the Future. Pearce, A.R., Ahn, Y.H., and Hanmi Global. (2012). Routledge, London, UK.

Sustainable Construction: Green Building Design and Delivery, 3rd Ed. Kibert, C.J. (2012). Wiley, New York, NY.

The HOK Guidebook to Sustainable Design, 3rd Ed. Odell, W. and Lazarus, M.A. (2015). Wiley, New York, NY.

The following list is a compilation of websites referenced in this module:

American Society of Heating, Refrigerating, and Air-Conditioning Engineers: **www.ashrae.org**

Arid Solutions Inc.: **www.aridsolutionsinc.com**

Carbon Footprint: **www.carbonfootprint.com**

Database of State Incentives for Renewables & Efficiency: **www.dsireusa.org**

Energy Star: **www.energystar.gov**

Forest Stewardship Council: **www.fsc.org**

Green Building Certification Institute: **www.gbci.org**

Green Building Initiative: **www.thegbi.org**

Green Globes: **www.greenglobes.com**

Green Seal: **www.greenseal.org**

Green-e: **www.green-e.org**

Habitat for Humanity: **www.habitat.org**

International Initiative for a Sustainable Built Environment: **www.iisbe.org**

Living Building Challenge: **www.living-future.org/lbc**

NAHB Research Center: **www.nahbrc.org**

Natural Capitalism: **www.natcap.org**

Office of the Federal Environmental Executive: **www.ofee.gov**

Refining Process: **www.myfootprint.org**

Sheet Metal and Air Conditioning Contractors' National Association: **www.smacna.org**

Smart Communities Network: **www.smartcommunities.ncat.org**

The Carpet and Rug Institute: **www.carpet-rug.org**

The PLANTS Database: **www.plants.usda.gov**

Unit Conversion: **www.convertunits.com**

U.S. Environmental Protection Agency: Clean Energy **www.epa.gov/cleanenergy**

U.S. Green Building Council: **www.usgbc.org**

WaterSense: **www.epa.gov/watersense**

Whole Building Design Guide: **www.wbdg.org**

Figure Credits

NOAA, *Figure 1*

NASA/NOAA GOES Project, Figure 2

Library of Congress, Prints & Photographs Division, photograph by Carol M. Highsmith (Reproduction Number, LC-DIG-highsm-13380), Figure 3

US Environmental Protection Agency, Figure 4A

US Geological Survey/photo by Robert Kamilli, Figure 4B

Deere & Company, Figures 4C, 78

US Geological Survey, Figures 4D, 67

Steve Hillebrand/USFWS, Figure 4E

USDA Forest Service/Frank Koch, Figure 4F

Henry Rose, Figure 5, SA06

Dr. Christine Fiori/Virginia Tech, Figures 6, 45

NASA, Figure 7

The City of Santa Monica, Figure 8

Sushil Shenoy/Virginia Tech, Figures 9, 16–18, 26, 29, 30-32, 35-44, 46, 51, 52, 56-61, 63, 65, 68, 72–77, 79, SA03, SA14, SA23, SA24

US Energy Information Administration (Sept 2014), *Figure 10*

Dr. Annie Pearce/Virginia Tech, Figures 11, 24, 25B, *27*, *48*, 49B, 53, *54*, *62*, *69–71* SA01, SA05, SA13, SA16, SA17

US Department of Energy Office of Energy Efficiency and Renewable Energy, Figure 15

Jamie Carroll/NCCER, Figures 19, 64

Waterless Co., Figure 20

Brac Systems, Figure 22

Tim Davis/NCCER, Figure 25A, SA07, SA22

Courtesy of the Center for Resource Solutions (**www.green-e.org**), Figure 28

Courtesy of USDA_NRCS, Figure 33A

Elizabeth Westfall, Figure 33B

Image provided by Tate Access Floors, Inc., Figure 47

Tim Dean, Figure 49A, SA25

Topaz Publications, Inc., *Figure 50*, SA02

© US Green Building Council, Figure 55, Appendix A

Arid Solutions Inc.com, Figure 66

Ren Solutions, Figure 80

USDA Forest Service, SA04

US Army photo by Todd Plain, SA08

@esperng (Foap.com), SA09

Pratik Doshi/Virginia Tech, SA11, SA18

Joshua Winchell/USFWS, SA12

NPS photo by Ed Austin/Herb Jones, SA15

US Army photo by Capt. Michael N. Meyer, SA19

Library of Congress, Prints & Photographs Division, photograph by Carol M. Highsmith (Reproduction Number, LC-DIG-highsm-13019), SA20

Library of Congress, Prints & Photographs Division, photograph by Carol M. Highsmith (Reproduction Number, LC-DIG-highsm-12111), SA21

Section Review Answer Key

Answer	Section Reference	Objective
Section One		
1. d	1.0.0	1a
2. b	1.1.2	1b
3. a	1.2.1	1c
Section Two		
1. c	2.0.0	2a
2 .b	2.2.3	2b
3.c	2.3.0	2c
4. c	2.4.1	2d
5. a	2.5.1	2e
6. b	2.6.1	2f
7. c	2.7.4	2g
Section Three		
1. b	3.0.0	3a
2. b	3.2.2	3b
3. d	3.3.0	3c
4. a	3.4.4	3d

NCCER CURRICULA — USER UPDATE

NCCER makes every effort to keep its textbooks up-to-date and free of technical errors. We appreciate your help in this process. If you find an error, a typographical mistake, or an inaccuracy in NCCER's curricula, please fill out this form (or a photocopy), or complete the online form at **www.nccer.org/olf**. Be sure to include the exact module ID number, page number, a detailed description, and your recommended correction. Your input will be brought to the attention of the Authoring Team. Thank you for your assistance.

Instructors – If you have an idea for improving this textbook, or have found that additional materials were necessary to teach this module effectively, please let us know so that we may present your suggestions to the Authoring Team.

NCCER Product Development and Revision

13614 Progress Blvd., Alachua, FL 32615

Email: curriculum@nccer.org
Online: www.nccer.org/olf

❏ Trainee Guide   ❏ Lesson Plans   ❏ Exam   ❏ PowerPoints   Other _____

Craft / Level: _____ Copyright Date: _____

Module ID Number / Title: _____

Section Number(s): _____

Description: _____

_____

_____

_____

_____

Recommended Correction: _____

_____

_____

_____

_____

Your Name: _____

Address: _____

_____

Email: _____ Phone: _____

Index

Heating, ventilation, and air conditioning (HVAC) systems. *See also* Ventilation
   best management practices (BMPs), 57–58
   geothermal, 57–58
   LEED documentation, 86
   LEED Green Building Rating System goals, 81
   optimizing energy use, 38, 39
Heating, ventilation, and air conditioning (HVAC) technicians, green, 104
Heat islands, 67, 75, 76, 97
Heat recovery ventilators (HRVs), 37
Heavy metals, 6
Hot water heating, 16, 19, 39
Household impacts inventory worksheet, 7–8, 108–109
Housekeeping practices, 55–56, 81
House size, increase in, 3
HRVs. *See* Heat recovery ventilators (HRVs)
Humidity levels in indoor environments, 55, 56, 57
Hurricanes, 3
HVAC. *See* Heating, ventilation, and air conditioning (HVAC) systems
Hybrid vehicle, 1, 17, 97
Hydrochlorofluorocarbons (HCFCs), 67, 76, 97
Hydrologic cycle, 23, 31, 97
Hydropower, 15

I

IAQ. *See* Indoor air quality (IAQ)
ICF. *See* Insulating concrete form (ICF)
ID. *See* Innovation in design (ID)
ID+C. *See* LEED for Interior Design and Construction (ID+C)
Indoor air quality (IAQ)
   best management practices, 54–58, 60
   construction materials impacts, 54–55
   defined, 23, 97
   finishes and, 55, 81
   flushout period, 56
   furnishings and, 54
   HVAC systems and, 57–58, 81, 86
   landscaping impacts, 54–55
   LEED documentation requirements, 87
   LEED Green Building Rating System goals, 78, 81–83
   maintenance for, 55
   poor, causes of, 25
   segregation, 55–57
   sequencing for, 55
   Sick Building Syndrome, 24, 25, 53, 99
   smoking facilities, 78, 81, 82
   using natural forces, 57–58
   ventilation and, 55–57, 58, 60
Indoor environmental quality (EQ)
   best management practices (BMPs), 28, 52–59
   built environment impact on, 26, 27
   category overview, 28
   climate-controlled buildings, 53
   color and, 57, 59
   daylighting, 57, 78, 81, 83
   landscaping impacts, 48, 54–55
   LEED performance goals and requirements for, 70
   problems, preventing, 54–55
   user controls, 58–59, 81
   using natural forces, 57–58
Infiltration, 23, 37, 97
Innovation in design (ID), 70
Insulating concrete form (ICF), 23, 47, 97
Integrative design, 23, 61, 97

Integrative process (IP), 69
IP. *See* Integrative process (IP)

J

Just-in-time delivery, 23, 50, 55, 97

L

Landfills, 23, 53, 98
Landscaping
   best management practices, 27, 29–31, 33
   ecosystem restoration, 31
   green roofs, 48
   indoor environment impacts, 29, 54–55
   LEED Green Building Rating System, 74–77
   LEED performance goals and requirements for, 78
   water conservation systems, 78
Leadership in Energy and Environmental Design (LEED), 67, 68, 97
Lean construction methods, 50
LED. *See* Light-emitting diode (LED) lamp
LEED. *See* Leadership in Energy and Environmental Design (LEED)
LEED accreditation, 74
LEED Accredited Professional (LEED AP), 85
LEED Certification
   accreditation vs., 74
   documentation requirements, 85–88
   levels, 70–71
   levels of, 70–71
   outside the US, 69
   process, 71–73
   project pitfalls
      lack of coordination with other trades, 88–89
      poor documentation, 87–88
      poor execution of credit requirements, 86–87
      poor planning, 86
LEED for Building Design + Construction (BD+C), 69, 70, 106–107
LEED for Building Operations and Maintenance (O+M), 70
LEED for Homes, 68
LEED for Interior Design and Construction (ID+C), 70
LEED for Neighborhood Development (ND), 70, 74
LEED for New Construction, 69
LEED for Schools, 83
LEED Gold standard, 68, 69, 71
LEED Minimum Program Requirements (MPRs), 70
LEED Online documentation system, 71
LEED Pilot Credit Library, 85
Leed Platinum standard, 71
LEED Rating System
   basis for, 68
   defined, 67
   function, 68
   goals
      by category, 69–70
      costs, 86–87
      healthy and productive living environments, 81–83
      promoting sustainable behavior, 77–78
      protecting and restoring the surrounding environment, 74–77
      resource conservation, 78–80
      site selection, 74
   requirements by category, 69–70
   structure, 69–70
   types of, 70–71
LEED Silver standard, 42, 71
LEED standards, 42, 54, 58–69, 71
Leverage points, 12–14